Health Risk and Exposure Assessment for Ozone

Second External Review Draft

Chapter 7, 8 and 9 Appendices

DISCLAIMER

This draft document has been prepared by staff from the Risk and Benefits Group, Health and Environmental Impacts Division, Office of Air Quality Planning and Standards, U.S. Environmental Protection Agency. Any findings and conclusions are those of the authors and do not necessarily reflect the views of the Agency. This draft document is being circulated to facilitate discussion with the Clean Air Scientific Advisory Committee to inform the EPA's consideration of the ozone National Ambient Air Quality Standards.

This information is distributed for the purposes of pre-dissemination peer review under applicable information quality guidelines. It has not been formally disseminated by EPA. It does not represent and should not be construed to represent any Agency determination or policy.

Questions related to this preliminary draft document should be addressed to Dr. Bryan Hubbell, U.S. Environmental Protection Agency, Office of Air Quality Planning and Standards, C539-07, Research Triangle Park, North Carolina 27711 (email: hubbell.bryan@epa.gov).

EPA-452/P-14-004e
February 2014

Health Risk and Exposure Assessment for Ozone
Second External Review Draft
Chapter 7, 8 and 9 Appendices

U.S. Environmental Protection Agency
Office of Air and Radiation
Office of Air Quality Planning and Standards
Health and Environmental Impacts Division
Risk and Benefits Group
Research Triangle Park, North Carolina 27711

Appendix 7A. Detailed Information on Effect Estimates, Baseline Incidence and Demographic Data Used in the Epidemiological-Based Risk Assessment

Endpoint	Study	Urban study area	Study area template	Air metric	Risk assessment modeling period	Age range	Lag	Additional study details	Statistical Model	Effect estimate (Beta)	SE (effect estimate)[a]	Baseline incidence[b] 2007	2009	Population 2007	2009
Core Risk - short-term exposure-related all-cause mortality															
Mortality, All Cause	Smith et al., 2009	Atlanta, GA	CBSA	D8HourMax	March-October	0-99	distributed lag 0-6 d	-	log-linear	0.00029536	0.000291921	24,086	24,565	5,798,520	5,991,005
Mortality, All Cause	Smith et al., 2009	Baltimore, MD	CBSA	D8HourMean	April-October	0-99	distributed lag 0-6 d	-	log-linear	0.000515048	0.000329964	22,709	22,630	5,362,979	5,437,691
Mortality, All Cause	Smith et al., 2009	Boston, MA	CBSA	D8HourMean	April-September	0-99	distributed lag 0-6 d	-	log-linear	0.000681639	0.000342908	29,168	28,606	7,469,168	7,553,629
Mortality, All Cause	Smith et al., 2009	Cleveland, OH	CBSA	D8HourMean	April-October	0-99	distributed lag 0-6 d	-	log-linear	0.000596249	0.000314904	17,651	17,246	3,419,753	3,404,546
Mortality, All Cause	Smith et al., 2009	Denver, CO	CBSA	D8HourMean	March-September	0-99	distributed lag 0-6 d	-	log-linear	0.000351818	0.000356513	9,977	10,128	3,585,323	3,714,085
Mortality, All Cause	Smith et al., 2009	Detroit, MI	CBSA	D8HourMean	April-September	0-99	distributed lag 0-6 d	-	log-linear	0.001045932	0.000311744	21,796	21,387	5,703,138	5,620,925
Mortality, All Cause	Smith et al., 2009	Houston, TX	CBSA	D8HourMean	January-December	0-99	distributed lag 0-6 d	-	log-linear	0.000162925	0.000207509	35,544	36,135	6,145,152	6,437,742
Mortality, All Cause	Smith et al., 2009	Los Angeles, CA	CBSA	D8HourMean	January-December	0-99	distributed lag 0-6 d	-	log-linear	0.000273722	0.000157143	121,194	121,736	21,225,780	21,587,310
Mortality, All Cause	Smith et al., 2009	New York, NY	CBSA	D8HourMean	April-October	0-99	distributed lag 0-6 d	-	log-linear	0.001092475	0.000207428	67,939	66,898	16,024,400	16,202,260
Mortality, All Cause	Smith et al., 2009	Philadelphia, PA	CBSA	D8HourMean	April-October	0-99	distributed lag 0-6 d	-	log-linear	0.000624582	0.000284572	38,076	37,426	7,813,329	7,904,328
Mortality, All Cause	Smith et al., 2009	Sacramento, CA	CBSA	D8HourMean	January-December	0-99	distributed lag 0-6 d	-	log-linear	0.000569111	0.00031446	30,170	30,336	4,675,398	4,770,990
Mortality, All Cause	Smith et al., 2009	St. Louis, MO	CBSA	D8HourMean	April-October	0-99	distributed lag 0-6 d	-	log-linear	0.000544366	0.000342796	17,256	16,888	3,344,163	3,369,708
Core Risk - long-term exposure-related respiratory mortality															
Mortality, Respiratory	Jerrett et al., 2009	Atlanta, GA	CBSA	Seasonal-avg D1hrMax	April-September	30-99	NA	-	log-linear	0.003922071	0.001324866	3,803	3,893	3,283,262	3,419,286
Mortality, Respiratory	Jerrett et al., 2010	Baltimore, MD	CBSA	Seasonal-avg D1hrMax	April-September	30-99	NA	-	log-linear	0.003922071	0.001324866	3,970	3,952	3,195,786	3,255,696
Mortality, Respiratory	Jerrett et al., 2011	Boston, MA	CBSA	Seasonal-avg D1hrMax	April-September	30-99	NA	-	log-linear	0.003922071	0.001324866	6,466	6,328	4,562,351	4,631,833
Mortality, Respiratory	Jerrett et al., 2012	Cleveland, OH	CBSA	Seasonal-avg D1hrMax	April-September	30-99	NA	-	log-linear	0.003922071	0.001324866	2,947	2,873	2,105,949	2,107,957
Mortality, Respiratory	Jerrett et al., 2013	Denver, CO	CBSA	Seasonal-avg D1hrMax	April-September	30-99	NA	-	log-linear	0.003922071	0.001324866	2,287	2,324	2,055,105	2,137,319
Mortality, Respiratory	Jerrett et al., 2014	Detroit, MI	CBSA	Seasonal-avg D1hrMax	April-September	30-99	NA	-	log-linear	0.003922071	0.001324866	4,094	4,007	3,382,306	3,373,240
Mortality, Respiratory	Jerrett et al., 2015	Houston, TX	CBSA	Seasonal-avg D1hrMax	April-September	30-99	NA	-	log-linear	0.003922071	0.001324866	3,317	3,370	3,359,712	3,529,238
Mortality, Respiratory	Jerrett et al., 2016	Los Angeles, CA	CBSA	Seasonal-avg D1hrMax	April-September	30-99	NA	-	log-linear	0.003922071	0.001324866	12,443	12,529	11,723,570	12,038,790
Mortality, Respiratory	Jerrett et al., 2017	New York, NY	CBSA	Seasonal-avg D1hrMax	April-September	30-99	NA	-	log-linear	0.003922071	0.001324866	10,779	10,600	9,670,019	9,817,407
Mortality, Respiratory	Jerrett et al., 2018	Philadelphia, PA	CBSA	Seasonal-avg D1hrMax	April-September	30-99	NA	PM2.5	log-linear	0.003922071	0.001324866	6,747	6,620	4,647,403	4,726,359
Mortality, Respiratory	Jerrett et al., 2019	Sacramento, CA	CBSA	Seasonal-avg D1hrMax	April-September	30-99	NA	-	log-linear	0.003922071	0.001324866	3,814	3,835	2,695,086	2,765,834
Mortality, Respiratory	Jerrett et al., 2020	St. Louis, MO	CBSA	Seasonal-avg D1hrMax	April-September	30-99	NA	-	log-linear	0.003922071	0.001324866	3,143	3,072	1,998,779	2,028,727
Core Risk - short-term exposure-related respiratory morbidity															
HA, All Respiratory	Katsouyanni et al., 2009	Detroit, MI	CBSA	D1HourMax	June-August	65-99	average of lag 0 and lag 1	penalized splines	log-linear	0.00056	0.000352041	8,291	8,519	687,389	713,374
HA, All Respiratory	Katsouyanni et al., 2009	Detroit, MI	CBSA	D1HourMax	June-August	65-99	average of lag 0 and lag 1	natural splines	log-linear	0.00054	0.000357143	8,291	8,519	687,389	713,374
HA, Asthma	Silverman and Ito, 2010	New York, NY	CBSA	D8HourMax	April-October	6-18	average of lag 0 and lag 1	-	log-linear	0.007906969	0.0037862	1,463	1,453	2,804,642	2,787,619
HA, Asthma	Silverman and Ito, 2010	New York, NY	CBSA	D8HourMax	April-October	6-18	average of lag 0 and lag 1	PM2.5	log-linear	0.005555347	0.003692645	1,463	1,453	2,804,642	2,787,619
HA, Chronic Lung Disease	Lin et al. (a), 2008	New York, NY	CBSA	D1HourMax	April-October	0-17	Lag 2 d	-	log-linear	0.000076087	0.000163043	3,706	3,667	3,823,944	3,786,262
HA, All Respiratory	Linn et al., 2000	Los Angeles, CA	CBSA	D24HourMean	June-August	30-99	Lag 0d	-	log-linear	0.0006	0.0007	32,087	33,749	11,723,570	12,038,790
HA, Chronic Lung Disease (less Asthma)	Medina-Ramon et al, 2006	Atlanta, GA	CBSA	D8HourMean	June-August	65-99	distributed lag 0-1 d	-	logistic	0.00054	0.00019898	2,646	2,883	505,741	554,624

Endpoint	Study	Urban study area	Study area template	Air metric	Risk assessment modeling period	Age range	Lag	Additional study details	Statistical Model	Effect estimate (Beta)	SE (effect estimate)[a]	Baseline incidence[b] 2007	Baseline incidence[b] 2009	Population 2007	Population 2009
HA, Chronic Lung Disease (less Asthma)	Medina-Ramon et al, 2006	Baltimore, MD	CBSA	D8HourMean	June-August	65-99	distributed lag 0 1 d	-	logistic	0.00054	0.00019898	2,958	3,086	621,817	654,073
HA, Chronic Lung Disease (less Asthma)	Medina-Ramon et al, 2006	Boston, MA	CBSA	D8HourMean	June-August	65-99	distributed lag 0 1 d	-	logistic	0.00054	0.00019898	4,502	4,618	975,770	1,009,556
HA, Chronic Lung Disease (less Asthma)	Medina-Ramon et al, 2006	Cleveland, OH	CBSA	D8HourMean	June-August	65-99	distributed lag 0 1 d	-	logistic	0.00054	0.00019898	2,569	2,619	495,715	507,939
HA, Chronic Lung Disease (less Asthma)	Medina-Ramon et al, 2006	Denver, CO	CBSA	D8HourMean	June-August	65-99	distributed lag 0 1 d	-	logistic	0.00054	0.00019898	923	987	337,427	365,450
HA, Chronic Lung Disease (less Asthma)	Medina-Ramon et al, 2006	Detroit, MI	CBSA	D8HourMean	June-August	65-99	distributed lag 0 1 d	-	logistic	0.00054	0.00019898	3,643	3,740	687,389	713,374
HA, Chronic Lung Disease (less Asthma)	Medina-Ramon et al, 2006	Houston, TX	CBSA	D8HourMean	June-August	65-99	distributed lag 0 1 d	-	logistic	0.00054	0.00019898	3,238	3,452	536,446	576,473
HA, Chronic Lung Disease (less Asthma)	Medina-Ramon et al, 2006	Los Angeles, CA	CBSA	D8HourMean	June-August	65-99	distributed lag 0 1 d	-	logistic	0.00054	0.00019898	6,741	7,165	2,183,030	2,301,532
HA, Chronic Lung Disease (less Asthma)	Medina-Ramon et al, 2006	New York, NY	CBSA	D8HourMean	June-August	65-99	distributed lag 0 1 d	-	logistic	0.00054	0.00019898	7,850	8,058	2,052,957	2,120,805
HA, Chronic Lung Disease (less Asthma)	Medina-Ramon et al, 2006	Philadelphia, PA	CBSA	D8HourMean	June-August	65-99	distributed lag 0 1 d	-	logistic	0.00054	0.00019898	5,173	5,315	1,023,602	1,059,325
HA, Chronic Lung Disease (less Asthma)	Medina-Ramon et al, 2006	Sacramento, CA	CBSA	D8HourMean	June-August	65-99	distributed lag 0 1 d	-	logistic	0.00054	0.00019898	1,385	1,475	536,631	569,298
HA, Chronic Lung Disease (less Asthma)	Medina-Ramon et al, 2006	St. Louis, MO	CBSA	D8HourMean	June-August	65-99	distributed lag 0 1 d	-	logistic	0.00054	0.00019898	2,048	2,099	442,691	456,212
Emergency Room Visits, Respiratory	Strickland et al., 2010	Atlanta, GA	Atlanta, GA	D8HourMax	March-October (8)	5-17	distributed lag 0 7 d	-	log-linear	0.004786368	0.000760164	38,242	39,464	1,105,830	1,141,180
Emergency Room Visits, Respiratory	Strickland et al., 2010	Atlanta, GA	Atlanta, GA	D8HourMax	March-October (8)	5-17	average of lags 0-2	-	log-linear	0.002699013	0.00064564	38,242	39,464	1,105,830	1,141,180
Emergency Room Visits, Respiratory	Tolbert et al., 2007	Atlanta, GA	Atlanta, GA	D8HourMax	March-October (8)	0-99	average of lags 0-2	-	log-linear	0.001286007	0.000206235	140,690	145,038	5,798,520	5,991,005
Emergency Room Visits, Respiratory	Tolbert et al., 2007	Atlanta, GA	Atlanta, GA	D8HourMax	March-October (8)	0-99	average of lags 0-2	CO	log-linear	0.0011408	0.000228328	140,690	145,038	5,798,520	5,991,005
Emergency Room Visits, Respiratory	Tolbert et al., 2007	Atlanta, GA	Atlanta, GA	D8HourMax	March-October (8)	0-99	average of lags 0-2	NO2	log-linear	0.001028713	0.000250581	140,690	145,038	5,798,520	5,991,005
Emergency Room Visits, Respiratory	Tolbert et al., 2007	Atlanta, GA	Atlanta, GA	D8HourMax	March-October (8)	0-99	average of lags 0-2	PM10	log-linear	0.000803233	0.000266964	140,690	145,038	5,798,520	5,991,005
Emergency Room Visits, Respiratory	Tolbert et al., 2007	Atlanta, GA	Atlanta, GA	D8HourMax	March-October (8)	0-99	average of lags 0-2	PM10, NO2	log-linear	0.000774925	0.000267224	140,690	145,038	5,798,520	5,991,005
Emergency Room Visits, Respiratory	Darrow et al., 2011	Atlanta, GA	Atlanta, GA	D8HourMax	March-October (8)	0-99	Lag 1d	-	log-linear	0.000685212	0.000138467	140,690	145,038	5,798,520	5,991,005
Emergency Room Visits, Asthma	Ito et al., 2007	New York, NY	New York, NY	D8HourMax	April-October (7)	0-99	average of lag 0 and lag 1	-	log-linear	0.005213389	0.00090866	45,290	45,547	16,024,400	16,202,260
Emergency Room Visits, Asthma	Ito et al., 2007	New York, NY	New York, NY	D8HourMax	April-October (7)	0-99	average of lag 0 and lag 1	PM2.5	log-linear	0.00397574	0.000978924	45,290	45,547	16,024,400	16,202,260
Emergency Room Visits, Asthma	Ito et al., 2007	New York, NY	New York, NY	D8HourMax	April-October (7)	0-99	average of lag 0 and lag 1	NO2	log-linear	0.003233689	0.00093586	45,290	45,547	16,024,400	16,202,260
Emergency Room Visits, Asthma	Ito et al., 2007	New York, NY	New York, NY	D8HourMax	April-October (7)	0-99	average of lag 0 and lag 1	CO	log-linear	0.005543699	0.000893945	45,290	45,547	16,024,400	16,202,260
Emergency Room Visits, Asthma	Ito et al., 2007	New York, NY	New York, NY	D8HourMax	April-October (7)	0-99	average of lag 0 and lag 1	SO2	log-linear	0.004114984	0.000922644	45,290	45,547	16,024,400	16,202,260
Asthma Exacerbation, Chest Tightness	Gent et al., 2003	Boston, MA	Boston, MA	D1HourMax	April-September (6)	0-12	Lag 1d	-	logistic	0.00076087	0.000000215	235,224	233,053	1,189,925	1,177,644
Asthma Exacerbation, Chest Tightness	Gent et al., 2003	Boston, MA	Boston, MA	D8HourMax	April-September (6)	0-12	Lag 1d	-	logistic	0.005703579	0.002021676	294,030	291,316	1,189,925	1,177,644
Asthma Exacerbation, Chest Tightness	Gent et al., 2003	Boston, MA	Boston, MA	D1HourMax	April-September (6)	0-12	Lag 1d	PM2.5	logistic	0.007705248	0.002266587	235,224	233,053	1,189,925	1,177,644
Asthma Exacerbation, Chest Tightness	Gent et al., 2003	Boston, MA	Boston, MA	D1HourMax	April-September (6)	0-12	Lag 1d	PM2.5	logistic	0.007013137	0.002273393	294,030	291,316	1,189,925	1,177,644

Endpoint	Study	Urban study area	Study area template	Air metric	Risk assessment modeling period	Age range	Lag	Additional study details	Statistical Model	Effect estimate (Beta)	SE (effect estimate)[a]	Baseline incidence[b] 2007	Baseline incidence[b] 2009	Population 2007	Population 2009
Asthma Exacerbation, Shortness of Breath	Gent et al., 2003	Boston, MA	Boston, MA	D1HourMax	April-September (6)	0-12	Lag 1d	-	logistic	0.003977017	0.001794699	235,224	233,053	1,189,925	1,177,644
Asthma Exacerbation, Shortness of Breath	Gent et al., 2003	Boston, MA	Boston, MA	D8HourMax	April-September (6)	0-12	Lag 1d	-	logistic	0.005247285	0.002180887	235,224	233,053	1,189,925	1,177,644
Asthma Exacerbation, Wheeze	Gent et al., 2003	Boston, MA	Boston, MA	D1HourMax	April-September (6)	0-12	Lag 0d	PM2.5	logistic	0.006002092	0.002022527	548,857	543,790	1,189,925	1,177,644

Sensitivity Analysis - short-term exposure-related all-cause mortality

Endpoint	Study	Urban study area	Study area template	Air metric	Risk assessment modeling period	Age range	Lag	Additional study details	Statistical Model	Effect estimate (Beta)	SE (effect estimate)[a]	Baseline incidence[b] 2007	Baseline incidence[b] 2009	Population 2007	Population 2009
Mortality, All Cause	Smith et al., 2009	Atlanta, GA	Epi study based	D8HourMax	March-October	0-99	distributed lag 0-6 d	-	log-linear	0.00029536	0.000291921		16,524		4,462,663
Mortality, All Cause	Smith et al., 2009	Baltimore, MD	Epi study based	D8HourMean	April-October	0-99	distributed lag 0-6 d	-	log-linear	0.000515048	0.000329964		11,341		1,953,317
Mortality, All Cause	Smith et al., 2009	Boston, MA	Epi study based	D8HourMean	April-September	0-99	distributed lag 0-6 d	-	log-linear	0.000681639	0.000342908		14,399		3,609,318
Mortality, All Cause	Smith et al., 2009	Cleveland, OH	Epi study based	D8HourMean	April-October	0-99	distributed lag 0-6 d	-	log-linear	0.000596249	0.000314904		15,402		2,786,348
Mortality, All Cause	Smith et al., 2009	Denver, CO	Epi study based	D8HourMean	March-September	0-99	distributed lag 0-6 d	-	log-linear	0.000351818	0.000356513		9,093		2,737,299
Mortality, All Cause	Smith et al., 2009	Detroit, MI	Epi study based	D8HourMax	April-September	0-99	distributed lag 0-6 d	-	log-linear	0.001045932	0.000311744	SA completed for 2009	19,846	SA completed for 2009	4,377,305
Mortality, All Cause	Smith et al., 2009	Houston, TX	Epi study based	D8HourMean	January-December	0-99	distributed lag 0-6 d	-	log-linear	0.000162925	0.000207509		29,179		5,769,285
Mortality, All Cause	Smith et al., 2009	Los Angeles, CA	Epi study based	D8HourMean	January-December	0-99	distributed lag 0-6 d	-	log-linear	0.000273722	0.000157143		92,186		16,403,420
Mortality, All Cause	Smith et al., 2009	New York, NY	Epi study based	D8HourMean	April-October	0-99	distributed lag 0-6 d	-	log-linear	0.001092475	0.000207428		50,341		13,239,830
Mortality, All Cause	Smith et al., 2009	Philadelphia, PA	Epi study based	D8HourMean	April-October	0-99	distributed lag 0-6 d	-	log-linear	0.000624582	0.000284572		27,057		4,737,330
Mortality, All Cause	Smith et al., 2009	Sacramento, CA	Epi study based	D8HourMean	January-December	0-99	distributed lag 0-6 d	-	log-linear	0.000569111	0.00031446		29,479		4,341,150
Mortality, All Cause	Smith et al., 2009	St. Louis, MO	Epi study based	D8HourMean	April-October	0-99	distributed lag 0-6 d	-	log-linear	0.000544366	0.000342796		11,625		1,855,249
Mortality, All Cause	Smith et al., 2009	Atlanta, GA	CBSA	D8HourMax	March-October	0-99	distributed lag 0-6 d	Regional Bayes-based	log-linear	0.000260274	0.000235902		24,565		5,991,005
Mortality, All Cause	Smith et al., 2009	Baltimore, MD	CBSA	D8HourMean	April-October	0-99	distributed lag 0-6 d	Regional Bayes-based	log-linear	0.000939893	0.000282919		22,630		5,437,691
Mortality, All Cause	Smith et al., 2009	Boston, MA	CBSA	D8HourMean	April-September	0-99	distributed lag 0-6 d	Regional Bayes-based	log-linear	0.000882699	0.000300367		28,606		7,553,629
Mortality, All Cause	Smith et al., 2009	Cleveland, OH	CBSA	D8HourMean	April-October	0-99	distributed lag 0-6 d	Regional Bayes-based	log-linear	0.000678936	0.000263728		17,246		3,404,546
Mortality, All Cause	Smith et al., 2009	Denver, CO	CBSA	D8HourMean	March-September	0-99	distributed lag 0-6 d	Regional Bayes-based	log-linear	0.0000293	0.000350178		10,128		3,714,085
Mortality, All Cause	Smith et al., 2009	Detroit, MI	CBSA	D8HourMean	April-September	0-99	distributed lag 0-6 d	Regional Bayes-based	log-linear	0.000715864	0.000262244	SA completed for 2009	21,387	SA completed for 2009	5,620,925
Mortality, All Cause	Smith et al., 2009	Houston, TX	CBSA	D8HourMean	January-December	0-99	distributed lag 0-6 d	Regional Bayes-based	log-linear	0.000422972	0.000182484		36,135		6,437,742
Mortality, All Cause	Smith et al., 2009	Los Angeles, CA	CBSA	D8HourMean	January-December	0-99	distributed lag 0-6 d	Regional Bayes-based	log-linear	0.000198781	0.000150979		121,736		21,587,310
Mortality, All Cause	Smith et al., 2009	New York, NY	CBSA	D8HourMean	April-October	0-99	distributed lag 0-6 d	Regional Bayes-based	log-linear	0.001122295	0.000180774		66,898		16,202,260
Mortality, All Cause	Smith et al., 2009	Philadelphia, PA	CBSA	D8HourMean	April-October	0-99	distributed lag 0-6 d	Regional Bayes-based	log-linear	0.001025996	0.000323469		37,426		7,904,328
Mortality, All Cause	Smith et al., 2009	Sacramento, CA	CBSA	D8HourMean	January-December	0-99	distributed lag 0-6 d	Regional Bayes-based	log-linear	0.000107022	0.000323012		30,336		4,770,990
Mortality, All Cause	Smith et al., 2009	St. Louis, MO	CBSA	D8HourMean	March-September	0-99	distributed lag 0-6 d	Regional Bayes-based	log-linear	0.000675448	0.00028		16,888		3,369,708
Mortality, All Cause	Smith et al., 2009	Atlanta, GA	CBSA	D8HourMax	March-October	0-99	distributed lag 0-6 d	PM10	log-linear	0.000118303	0.000545629		24,565		5,991,005
Mortality, All Cause	Smith et al., 2009	Baltimore, MD	CBSA	D8HourMean	April-October	0-99	distributed lag 0-6 d	PM10	log-linear	0.000472682	0.000530993		22,630		5,437,691
Mortality, All Cause	Smith et al., 2009	Boston, MA	CBSA	D8HourMean	April-September	0-99	distributed lag 0-6 d	PM10	log-linear	0.000159064	0.000575152		28,606		7,553,629
Mortality, All Cause	Smith et al., 2009	Cleveland, OH	CBSA	D8HourMean	April-October	0-99	distributed lag 0-6 d	PM10	log-linear	0.000462588	0.00043506		17,246		3,404,546
Mortality, All Cause	Smith et al., 2009	Denver, CO	CBSA	D8HourMean	March-September	0-99	distributed lag 0-6 d	PM10	log-linear	-0.0000383	0.000526263		10,128		3,714,085
Mortality, All Cause	Smith et al., 2009	Detroit, MI	CBSA	D8HourMean	April-September	0-99	distributed lag 0-6 d	PM10	log-linear	0.000286042	0.000406606	SA completed for 2009	21,387	SA completed for 2009	5,620,925
Mortality, All Cause	Smith et al., 2009	Houston, TX	CBSA	D8HourMean	January-December	0-99	distributed lag 0-6 d	PM10	log-linear	0.000631017	0.000362269		36,135		6,437,742

Endpoint	Study	Urban study area	Study area template	Air metric	Risk assessment modeling period	Age range	Lag	Additional study details	Statistical Model	Effect estimate (Beta)	SE (effect estimate)[a]	Baseline incidence[b] 2007	Baseline incidence[b] 2009	Population 2007	Population 2009
Mortality, All Cause	Smith et al., 2009	Los Angeles, CA	CBSA	D8HourMean	January-December	0-99	distributed lag 0 6 d	PM10	log-linear	0.0000524	0.000347345		121,736		21,587,310
Mortality, All Cause	Smith et al., 2009	New York, NY	CBSA	D8HourMean	April-October	0-99	distributed lag 0 6 d	PM10	log-linear	0.000440658	0.000390403		66,898		16,202,260
Mortality, All Cause	Smith et al., 2009	Philadelphia, PA	CBSA	D8HourMean	April-October	0-99	distributed lag 0 6 d	PM10	log-linear	0.000544544	0.000518628		37,426		7,904,328
Mortality, All Cause	Smith et al., 2009	Sacramento, CA	CBSA	D8HourMean	January-December	0-99	distributed lag 0 6 d	PM10	log-linear	0.000280547	0.000543375		30,336		4,770,990
Mortality, All Cause	Smith et al., 2009	St. Louis, MO	CBSA	D8HourMean	April-October	0-99	distributed lag 0 6 d	PM10	log-linear	0.000360175	0.000581296		16,888		3,369,708
Mortality, All Cause	Zanobetti and Schwartz (b), 2008	Atlanta, GA	CBSA	D8HourMean	June-August	0-99	distributed lag 0 3 d	-	log-linear	0.00029536	0.0002288562	SA completed for 2009	10,119	SA completed for 2009	5,991,005
Mortality, All Cause	Zanobetti and Schwartz (b), 2008	Baltimore, MD	CBSA	D8HourMean	June-August	0-99	distributed lag 0 3 d	-	log-linear	0.000515048	0.00031402		10,408		5,437,691
Mortality, All Cause	Zanobetti and Schwartz (b), 2008	Boston, MA	CBSA	D8HourMean	June-August	0-99	distributed lag 0 3 d	-	log-linear	0.000681639	0.000328429		15,160		7,553,629
Mortality, All Cause	Zanobetti and Schwartz (b), 2008	Cleveland, OH	CBSA	D8HourMean	June-August	0-99	distributed lag 0 3 d	-	log-linear	0.000596249	0.000354552		7,808		3,404,546
Mortality, All Cause	Zanobetti and Schwartz (b), 2008	Denver, CO	CBSA	D8HourMean	June-August	0-99	distributed lag 0 3 d	-	log-linear	0.000351818	0.000408829		4,953		3,714,085
Mortality, All Cause	Zanobetti and Schwartz (b), 2008	Detroit, MI	CBSA	D8HourMean	June-August	0-99	distributed lag 0 3 d	-	log-linear	0.001045932	0.000344122		11,430		5,620,925
Mortality, All Cause	Zanobetti and Schwartz (b), 2008	Houston, TX	CBSA	D8HourMean	June-August	0-99	distributed lag 0 3 d	-	log-linear	0.000162925	0.000262836		10,132		6,437,742
Mortality, All Cause	Zanobetti and Schwartz (b), 2008	Los Angeles, CA	CBSA	D8HourMean	June-August	0-99	distributed lag 0 3 d	-	log-linear	0.000273722	0.000213402		32,840		21,587,310
Mortality, All Cause	Zanobetti and Schwartz (b), 2008	New York, NY	CBSA	D8HourMean	June-August	0-99	distributed lag 0 3 d	-	log-linear	0.001092475	0.000235676		30,192		16,202,260
Mortality, All Cause	Zanobetti and Schwartz (b), 2008	Philadelphia, PA	CBSA	D8HourMean	June-August	0-99	distributed lag 0 3 d	-	log-linear	0.000624582	0.000314555		17,137		7,904,328
Mortality, All Cause	Zanobetti and Schwartz (b), 2008	Sacramento, CA	CBSA	D8HourMean	June-August	0-99	distributed lag 0 3 d	-	log-linear	0.000569111	0.000388525		8,295		4,770,990
Mortality, All Cause	Zanobetti and Schwartz (b), 2008	St. Louis, MO	CBSA	D8HourMean	June-August	0-99	distributed lag 0 3 d	-	log-linear	0.000544366	0.000333395		7,837		3,369,708

Sensitivity Analysis - long-term exposure-related respiratory mortality

Endpoint	Study	Urban study area	Study area template	Air metric	Risk assessment modeling period	Age range	Lag	Additional study details	Statistical Model	Effect estimate (Beta)	SE (effect estimate)[a]	Baseline incidence[b] 2007	Baseline incidence[b] 2009	Population 2007	Population 2009
Mortality, Respiratory	Jerrett et al., 2009	Atlanta, GA	CBSA	Seasonal-avg D1hrMax	April-September	30-99	NA	Reginoal	log-linear	0.0031923937	0.0027397	3,803	3,893	3,283,262	3,419,286
Mortality, Respiratory	Jerrett et al., 2010	Baltimore, MD	CBSA	Seasonal-avg D1hrMax	April-September	30-99	NA	Reginoal	log-linear	0.003853068	0.0027397	3,970	3,952	3,195,786	3,255,696
Mortality, Respiratory	Jerrett et al., 2011	Boston, MA	CBSA	Seasonal-avg D1hrMax	April-September	30-99	NA	Reginoal	log-linear	0.003853068	0.0027397	6,466	6,328	4,562,351	4,631,833
Mortality, Respiratory	Jerrett et al., 2012	Cleveland, OH	CBSA	Seasonal-avg D1hrMax	April-September	30-99	NA	Reginoal	log-linear	0.004604295	0.0027397	2,947	2,873	2,105,949	2,107,957
Mortality, Respiratory	Jerrett et al., 2013	Denver, CO	CBSA	Seasonal-avg D1hrMax	April-September	30-99	NA	Reginoal	log-linear	0.0031117797	0.0027397	2,287	2,324	2,055,105	2,137,319
Mortality, Respiratory	Jerrett et al., 2014	Detroit, MI	CBSA	Seasonal-avg D1hrMax	April-September	30-99	NA	Reginoal	log-linear	0.004604295	0.0027397	4,094	4,007	3,382,306	3,373,240
Mortality, Respiratory	Jerrett et al., 2015	Houston, TX	CBSA	Seasonal-avg D1hrMax	April-September	30-99	NA	Reginoal	log-linear	0.0031923937	0.0027397	3,317	3,370	3,359,712	3,529,238
Mortality, Respiratory	Jerrett et al., 2016	Los Angeles, CA	CBSA	Seasonal-avg D1hrMax	April-September	30-99	NA	Reginoal	log-linear	0.002767363	0.0027397	12,443	12,529	11,723,570	12,038,790
Mortality, Respiratory	Jerrett et al., 2017	New York, NY	CBSA	Seasonal-avg D1hrMax	April-September	30-99	NA	Reginoal	log-linear	0.003853068	0.0027397	10,779	10,600	9,670,019	9,817,407
Mortality, Respiratory	Jerrett et al., 2018	Philadelphia, PA	CBSA	Seasonal-avg D1hrMax	April-September	30-99	NA	Reginoal	log-linear	0.003853068	0.0027397	6,747	6,620	4,647,403	4,726,359
Mortality, Respiratory	Jerrett et al., 2019	Sacramento, CA	CBSA	Seasonal-avg D1hrMax	April-September	30-99	NA	Reginoal	log-linear	0.0031117797	0.0027397	3,814	3,835	2,695,086	2,765,834
Mortality, Respiratory	Jerrett et al., 2020	St. Louis, MO	CBSA	Seasonal-avg D1hrMax	April-September	30-99	NA	Reginoal	log-linear	0.004604295	0.0027397	3,143	3,072	1,998,779	2,028,727

a-all Beta distributions assumed to be normal

b-Gent et al., 2003 also usese the following prevalence rates 0.028 (wheeze), 0.015 (shortness of breath), 0.012 (chest tightness) (from study

Appendix 7-B – Detailed Summary of Core Risk Estimates

Table 7B-1 Core Short-Term Ozone-Attributable Mortality (2007) (incidence, percent of baseline mortality, incidence per 100,000) (Smith et al., 2009)

| | Air Quality Scenario | | | | | | | |
| | Absolute Ozone-Attributable Incidence | | | | Change in Ozone-Attributable Incidence | | | |
Study Area	Base	75ppb	70ppb	65ppb	60ppb	Base-75	75-70	75-65	75-60
Atlanta, GA	300	270	260	250	230	37	12	21	34
	(-430 - 1000)	(-370 - 890)	(-360 - 850)	(-340 - 820)	(-330 - 780)	(-51 - 120)	(-16 - 39)	(-30 - 72)	(-47 - 110)
Baltimore, MD	470	440	430	420	400	23	13	27	45
	(-260 - 1200)	(-250 - 1100)	(-240 - 1100)	(-230 - 1000)	(-220 - 1000)	(-12 - 57)	(-7 - 33)	(-15 - 68)	(-25 - 110)
Boston, MA	360	350	350	330	320	6	7	20	32
	(-510 - 1200)	(-500 - 1200)	(-490 - 1200)	(-470 - 1100)	(-460 - 1100)	(-8 - 19)	(-10 - 24)	(-28 - 67)	(-45 - 110)
Cleveland, OH	430	430	420	400	370	0	14	32	64
	(-41 - 890)	(-41 - 890)	(-40 - 860)	(-38 - 830)	(-35 - 760)	(0 - -1)	(-1 - 28)	(-3 - 67)	(-6 - 130)
Denver, CO	87	86	84	82	79	1	2	4	8
	(-290 - 440)	(-280 - 440)	(-280 - 430)	(-270 - 420)	(-260 - 400)	(-3 - 4)	(-6 - 10)	(-14 - 23)	(-25 - 40)
Detroit, MI	660	660	630	610	590	3	23	42	69
	(32 - 1300)	(32 - 1300)	(31 - 1200)	(30 - 1200)	(29 - 1100)	(0 - 5)	(1 - 44)	(2 - 81)	(3 - 130)
Houston, TX	640	680	680	670	660	-46	5	11	24
	(120 - 1100)	(130 - 1200)	(130 - 1200)	(120 - 1200)	(120 - 1200)	(-8 - -83)	(1 - 9)	(2 - 20)	(4 - 43)
Los Angeles, CA	1100	1300	1200	1200	1100	-180	43	87	160
	(-450 - 2600)	(-530 - 3000)	(-510 - 2900)	(-490 - 2800)	(-460 - 2600)	(77 - -450)	(-18 - 100)	(-36 - 210)	(-66 - 380)
New York, NY	3000	2800	2700	2200	2200	150	130	640	NA
	(1800 - 4100)	(1700 - 3900)	(1600 - 3700)	(1300 - 3100)	(1300 - 3100)	(90 - 210)	(80 - 190)	(380 - 890)	NA
Philadelphia, PA	1300	1200	1200	1200	1100	64	35	76	120
	(290 - 2300)	(270 - 2200)	(260 - 2100)	(260 - 2000)	(250 - 2000)	(14 - 110)	(8 - 62)	(17 - 140)	(25 - 210)
Sacramento, CA	380	370	360	350	340	11	7	13	23
	(-400 - 1100)	(-390 - 1100)	(-380 - 1100)	(-380 - 1100)	(-370 - 1000)	(-12 - 33)	(-7 - 20)	(-13 - 39)	(-24 - 70)
St. Louis, MO	460	430	410	390	370	28	18	39	60
	(-110 - 1000)	(-110 - 950)	(-100 - 910)	(-98 - 870)	(-93 - 820)	(-7 - 62)	(-5 - 41)	(-10 - 86)	(-15 - 130)

| | Air Quality Scenario | | | | | | | |
| | Percent of Baseline Incidence Attributable to Ozone | | | | Change in O$_3$-Attributable Risk | | | |
Study Area	Base	75ppb	70ppb	65ppb	60ppb	Base-75	75-70	75-65	75-60
Atlanta, GA	1.2	1.1	1.0	1.0	1.0	12	4	8	12
Baltimore, MD	2.0	1.9	1.9	1.8	1.7	5	3	6	10
Boston, MA	1.2	1.2	1.2	1.1	1.1	2	2	5	9
Cleveland, OH	2.4	2.4	2.4	2.3	2.1	0	3	7	14
Denver, CO	0.8	0.8	0.8	0.8	0.8	1	2	5	9
Detroit, MI	3.0	3.0	2.9	2.8	2.7	0	3	6	10
Houston, TX	1.8	1.9	1.9	1.9	1.9	-7	1	2	3
Los Angeles, CA	0.9	1.0	1.0	1.0	0.9	-17	3	7	13
New York, NY	4.3	4.1	3.9	3.2	3.2	5	5	22	NA
Philadelphia, PA	3.4	3.2	3.2	3.0	2.9	5	3	6	9
Sacramento, CA	1.2	1.2	1.2	1.2	1.1	3	2	3	6
St. Louis, MO	2.6	2.5	2.4	2.3	2.1	6	4	9	14

| | Air Quality Scenario | | | | | | | |
| | Ozone-Attributable Deaths per 100,000 | | | | Change in Ozone-Attributable Deaths per 100,000 | | | |
Study Area	Base	75ppb	70ppb	65ppb	60ppb	Base-75	75-70	75-65	75-60
Atlanta, GA	5.17	4.66	4.48	4.31	3.97	0.64	0.21	0.36	0.59
Baltimore, MD	8.76	8.20	8.02	7.83	7.46	0.43	0.24	0.50	0.84
Boston, MA	4.82	4.69	4.69	4.42	4.28	0.08	0.10	0.27	0.43
Cleveland, OH	12.57	12.57	12.28	11.70	10.82	-0.01	0.41	0.94	1.87
Denver, CO	2.43	2.40	2.34	2.29	2.20	0.02	0.05	0.12	0.21
Detroit, MI	11.57	11.57	11.05	10.70	10.35	0.05	0.40	0.74	1.21
Houston, TX	10.41	11.07	11.07	10.90	10.74	-0.75	0.08	0.18	0.39
Los Angeles, CA	5.18	6.12	5.65	5.65	5.18	-0.85	0.20	0.41	0.75
New York, NY	18.72	17.47	16.85	13.73	13.73	0.94	0.81	3.99	NA
Philadelphia, PA	16.64	15.36	15.36	15.36	14.08	0.82	0.45	0.97	1.54
Sacramento, CA	8.13	7.91	7.70	7.49	7.27	0.24	0.14	0.28	0.49
St. Louis, MO	13.76	12.86	12.26	11.66	11.06	0.84	0.54	1.17	1.79

NA: for NYC, the model-based adjustment methodology was unable to estimate ozone distributions which would meet the lower alternative standard level of 60 ppb.

Table 7B-2 Core Short-Term Ozone-Attributable Mortality (2009) (incidence, percent of baseline mortality, incidence per 100,000) (Smith et al., 2009)

Study Area	Air Quality Scenario								
	Absolute Ozone-Attributable Incidence					Change in Ozone-Attributable Incidence			
	Base	75ppb	70ppb	65ppb	60ppb	Base-75	75-70	75-65	75-60
Atlanta, GA	250 (-340 - 820)	240 (-340 - 800)	230 (-320 - 770)	230 (-310 - 750)	220 (-300 - 730)	5 (-6 - 15)	9 (-12 - 28)	16 (-22 - 54)	23 (-32 - 77)
Baltimore, MD	410 (-230 - 1000)	400 (-220 - 1000)	400 (-220 - 990)	390 (-210 - 970)	380 (-210 - 940)	6 (-3 - 15)	7 (-4 - 19)	17 (-10 - 44)	28 (-15 - 71)
Boston, MA	320 (-450 - 1100)	320 (-450 - 1100)	320 (-450 - 1100)	310 (-450 - 1100)	310 (-430 - 1000)	-2 (3 - -8)	-1 (2 - -4)	5 (-7 - 17)	14 (-19 - 47)
Cleveland, OH	400 (-38 - 820)	400 (-38 - 830)	390 (-37 - 800)	370 (-36 - 770)	350 (-34 - 730)	-5 (0 - -10)	12 (-1 - 24)	29 (-3 - 60)	49 (-5 - 100)
Denver, CO	83 (-270 - 420)	83 (-270 - 420)	83 (-270 - 420)	81 (-270 - 410)	76 (-250 - 390)	0 (1 - -2)	0 (-1 - 2)	2 (-6 - 10)	7 (-23 - 37)
Detroit, MI	580 (28 - 1100)	580 (28 - 1100)	600 (29 - 1200)	590 (28 - 1100)	560 (27 - 1100)	NA	-21 (-1 - -41)	-6 (0 - -12)	15 (1 - 30)
Houston, TX	640 (120 - 1200)	700 (130 - 1200)	700 (130 - 1300)	690 (130 - 1200)	680 (130 - 1200)	-55 (-10 - -99)	0 (0 - -1)	4 (1 - 7)	14 (3 - 26)
Los Angeles, CA	1100 (-470 - 2700)	1300 (-540 - 3100)	1200 (-520 - 3000)	1200 (-500 - 2900)	1100 (-470 - 2700)	-160 (68 - -400)	41 (-17 - 100)	89 (-37 - 210)	160 (-68 - 390)
New York, NY	2600 (1500 - 3600)	2600 (1600 - 3700)	2600 (1500 - 3600)	2200 (1300 - 3100)	2200 (1300 - 3100)	-77 (-46 - -110)	84 (50 - 120)	440 (260 - 610)	NA
Philadelphia, PA	1100 (240 - 1900)	1100 (240 - 2000)	1100 (240 - 1900)	1100 (230 - 1900)	1000 (230 - 1800)	-6 (-1 - -10)	19 (4 - 34)	44 (10 - 78)	69 (15 - 120)
Sacramento, CA	380 (-400 - 1100)	370 (-390 - 1100)	360 (-380 - 1100)	360 (-380 - 1100)	350 (-370 - 1000)	10 (-11 - 31)	6 (-7 - 19)	12 (-13 - 38)	21 (-22 - 64)
St. Louis, MO	380 (-96 - 850)	380 (-96 - 840)	370 (-94 - 830)	360 (-91 - 800)	350 (-87 - 770)	1 (0 - 3)	8 (-2 - 18)	21 (-5 - 46)	37 (-9 - 83)

Study Area	Air Quality Scenario								
	Percent of Baseline Incidence Attributable to Ozone					Change in O_3-Attributable Risk			
	Base	75ppb	70ppb	65ppb	60ppb	Base-75	75-70	75-65	75-60
Atlanta, GA	1.0	1.0	0.9	0.9	0.9	2	3	7	9
Baltimore, MD	1.8	1.8	1.7	1.7	1.7	1	2	4	7
Boston, MA	1.1	1.1	1.1	1.1	1.1	-1	0	2	4
Cleveland, OH	2.3	2.3	2.3	2.2	2.0	-1	3	7	12
Denver, CO	0.8	0.8	0.8	0.8	0.7	0	0	2	8
Detroit, MI	2.7	2.7	2.8	2.7	2.6	0	-4	-1	3
Houston, TX	1.8	1.9	1.9	1.9	1.9	-8	0	0	2
Los Angeles, CA	0.9	1.1	1.0	1.0	0.9	-15	3	7	13
New York, NY	3.8	3.9	3.8	3.3	3.3	-3	3	16	NA
Philadelphia, PA	2.9	3.0	2.9	2.8	2.8	-1	2	4	6
Sacramento, CA	1.2	1.2	1.2	1.2	1.1	3	2	3	6
St. Louis, MO	2.3	2.3	2.2	2.1	2.0	0	2	5	9

Study Area	Air Quality Scenario								
	Ozone-Attributable Deaths per 100,000					Change in Ozone-Attributable Deaths per 100,000			
	Base	75ppb	70ppb	65ppb	60ppb	Base-75	75-70	75-65	75-60
Atlanta, GA	4.17	4.01	3.84	3.84	3.67	0.08	0.14	0.27	0.38
Baltimore, MD	7.54	7.36	7.36	7.17	6.99	0.11	0.14	0.31	0.51
Boston, MA	4.24	4.24	4.24	4.10	4.10	-0.03	-0.01	0.07	0.19
Cleveland, OH	11.75	11.75	11.46	10.87	10.28	-0.14	0.35	0.85	1.44
Denver, CO	2.23	2.23	2.23	2.18	2.05	-0.01	0.01	0.05	0.19
Detroit, MI	10.32	10.32	10.67	10.50	9.96	NA	-0.37	-0.11	0.27
Houston, TX	9.94	10.87	10.87	10.72	10.56	-0.85	-0.01	0.06	0.22
Los Angeles, CA	5.10	6.02	5.56	5.56	5.10	-0.74	0.19	0.41	0.74
New York, NY	16.05	16.05	16.05	13.58	13.58	-0.48	0.52	2.72	NA
Philadelphia, PA	13.92	13.92	13.92	13.92	12.65	-0.07	0.24	0.56	0.87
Sacramento, CA	7.96	7.76	7.55	7.55	7.34	0.21	0.13	0.25	0.44
St. Louis, MO	11.28	11.28	10.98	10.68	10.39	0.04	0.24	0.62	1.10

NA: for NYC, the model-based adjustment methodology was unable to estimate ozone distributions which would meet the lower alternative standard level of 60 ppb.

Figure 7B-1 Core Short-Term Ozone-Attributable Mortality (2007) (heat map tables – absolute ozone-attributable incidence) (Smith et al., 2009)

Recent conditions

Study area	Daily 8hr Max Ozone Level (ppb)																Total
	0-5	5-10	10-15	15-20	20-25	25-30	30-35	35-40	40-45	45-50	50-55	55-60	60-65	65-70	70-75	>75	
Atlanta, GA	0	0	0	2	3	5	13	16	20	29	41	57	35	36	24	23	304
Baltimore, MD	0	0	0	2	7	19	18	48	39	55	58	56	35	65	38	25	465
Boston, MA	0	0	0	1	9	17	47	42	62	38	30	31	25	18	10	29	358
Cleveland, OH	0	0	1	3	7	23	45	48	73	55	53	40	37	25	10	11	431
Denver, CO	0	0	0	0	0	0	3	5	7	9	13	18	17	8	4	2	87
Detroit, MI	0	0	1	0	7	29	40	61	96	121	63	39	51	25	37	86	657
Houston, TX	0	2	8	23	48	69	87	83	72	57	60	33	30	32	31	4	638
Los Angeles, CA	0	0	6	23	55	63	115	97	159	140	135	131	59	33	26	29	1,070
New York, NY	0	0	0	41	81	147	295	478	284	389	266	199	193	231	178	171	2,953
Philadelphia, PA	0	0	2	7	21	53	85	94	159	131	151	158	94	126	116	103	1,299
Sacramento, CA	0	0	0	4	17	23	38	56	44	47	46	44	23	17	12	6	378
St. Louis, MO	0	0	1	2	3	11	21	42	60	41	72	59	40	31	28	44	457

Current Standard (75)

Study area	Daily 8hr Max Ozone Level (ppb)																Total
	0-5	5-10	10-15	15-20	20-25	25-30	30-35	35-40	40-45	45-50	50-55	55-60	60-65	65-70	70-75	>75	
Atlanta, GA	0	0	0	0	3	5	18	24	41	52	63	38	15	6	3	0	267
Baltimore, MD	0	0	0	1	1	11	22	43	84	71	69	73	44	12	9	3	443
Boston, MA	0	0	0	0	4	20	45	50	58	57	35	20	30	9	13	11	353
Cleveland, OH	0	0	0	1	5	14	40	65	89	81	43	40	31	12	10	0	431
Denver, CO	0	0	0	0	0	0	1	5	6	13	17	23	15	4	2	0	86
Detroit, MI	0	0	0	0	2	7	42	72	123	147	75	52	56	20	43	17	655
Houston, TX	0	0	0	0	17	49	126	146	148	95	50	49	3	0	0	0	683
Los Angeles, CA	0	0	0	0	0	0	0	17	340	445	388	44	13	5	0	0	1,253
New York, NY	0	0	0	0	21	98	297	544	741	475	364	233	39	0	0	0	2,812
Philadelphia, PA	0	0	0	2	0	34	62	156	213	236	209	165	101	42	9	10	1,238
Sacramento, CA	0	0	0	0	1	18	53	98	67	65	40	20	5	2	0	0	367
St. Louis, MO	0	0	0	1	3	7	18	65	66	76	74	47	29	29	12	3	430

Alternative Standard 70

Study area	Daily 8hr Max Ozone Level (ppb)																Total
	0-5	5-10	10-15	15-20	20-25	25-30	30-35	35-40	40-45	45-50	50-55	55-60	60-65	65-70	70-75	>75	
Atlanta, GA	0	0	0	0	2	9	19	27	52	64	52	20	7	3	0	0	255
Baltimore, MD	0	0	0	1	1	11	14	54	95	86	84	51	22	9	3	0	430
Boston, MA	0	0	0	0	4	19	48	60	55	55	37	28	14	12	6	7	346
Cleveland, OH	0	0	0	1	3	16	41	72	108	76	39	34	23	6	0	0	418
Denver, CO	0	0	0	0	0	0	0	5	7	16	26	23	6	3	0	0	84
Detroit, MI	0	0	0	0	0	6	42	83	152	143	63	69	29	30	16	0	633
Houston, TX	0	0	0	0	9	49	128	166	164	96	54	13	0	0	0	0	678
Los Angeles, CA	0	0	0	0	0	0	0	28	399	603	164	8	9	0	0	0	1,211
New York, NY	0	0	0	0	13	136	341	652	809	519	195	17	0	0	0	0	2,684
Philadelphia, PA	0	0	0	0	3	31	61	180	273	225	216	121	77	8	9	0	1,204
Sacramento, CA	0	0	0	0	0	15	55	106	79	69	21	14	0	2	0	0	361
St. Louis, MO	0	0	0	1	2	8	24	76	76	84	58	42	30	12	0	0	412

Alternative Standard 65

Study area	Daily 8hr Max Ozone Level (ppb)																Total
	0-5	5-10	10-15	15-20	20-25	25-30	30-35	35-40	40-45	45-50	50-55	55-60	60-65	65-70	70-75	>75	
Atlanta, GA	0	0	0	0	2	10	24	28	65	66	37	9	4	0	0	0	246
Baltimore, MD	0	0	0	1	1	10	22	65	99	85	83	37	11	3	0	0	417
Boston, MA	0	0	0	0	2	19	53	63	65	54	37	20	11	6	3	0	333
Cleveland, OH	0	0	0	0	3	18	55	92	104	67	36	18	6	0	0	0	400
Denver, CO	0	0	0	0	0	0	1	3	10	20	31	14	3	0	0	0	82
Detroit, MI	0	0	0	0	0	4	42	94	183	121	71	47	37	15	0	0	615
Houston, TX	0	0	0	0	5	42	141	183	176	81	45	0	0	0	0	0	672
Los Angeles, CA	0	0	0	0	0	0	0	105	519	479	48	12	4	0	0	0	1,167
New York, NY	0	0	0	0	37	604	618	920	13	0	0	0	0	0	0	0	2,193
Philadelphia, PA	0	0	0	0	3	32	61	194	308	201	85	8	8	0	0	0	1,165
Sacramento, CA	0	0	0	0	0	11	65	114	93	50	17	4	2	0	0	0	355
St. Louis, MO	0	0	0	0	3	8	36	77	85	93	47	35	8	0	0	0	393

Alternative Standard 60

Study area	Daily 8hr Max Ozone Level (ppb)																Total
	0-5	5-10	10-15	15-20	20-25	25-30	30-35	35-40	40-45	45-50	50-55	55-60	60-65	65-70	70-75	>75	
Atlanta, GA	0	0	0	0	2	13	25	49	64	58	19	3	0	0	0	0	233
Baltimore, MD	0	0	0	1	1	10	23	87	108	108	48	13	0	0	0	0	399
Boston, MA	0	0	0	0	1	21	68	51	93	45	21	12	5	3	0	0	321
Cleveland, OH	0	0	0	0	4	24	82	106	112	24	17	0	0	0	0	0	369
Denver, CO	0	0	0	0	0	0	1	3	13	31	27	4	0	0	0	0	79
Detroit, MI	0	0	0	0	0	2	49	134	176	128	59	39	0	0	0	0	588
Houston, TX	0	0	0	0	0	33	160	227	179	57	5	0	0	0	0	0	659
Los Angeles, CA	0	0	0	0	0	0	12	375	439	251	19	0	0	0	0	0	1,095
New York, NY								NA									
Philadelphia, PA	0	0	0	0	3	28	82	218	355	295	131	7	8	0	0	0	1,126
Sacramento, CA	0	0	0	0	0	9	74	134	88	29	9	1	0	0	0	0	344
St. Louis, MO	0	0	0	0	3	10	56	91	114	57	36	5	0	0	0	0	371

NA: for NYC, the model-based adjustment methodology was unable to estimate ozone distributions which would meet the lower alternative standard level of 60 ppb.

Figure 7B-2 Core Short-Term Ozone-Attributable Mortality (2007) (heat map tables – change in absolute ozone-attributable incidence) (Smith et al., 2009)

Note: negative values are risk increases, positive values are risk reductions

Decrease recent conditions to 75

Study area	Daily 8hr Max Ozone Level (ppb)																Total	Change in risk	
	0-5	5-10	10-15	15-20	20-25	25-30	30-35	35-40	40-45	45-50	50-55	55-60	60-65	65-70	70-75	>75		Inc.	Dec.
Atlanta, GA	0	0	0	-1	0	0	0	0	1	2	5	8	6	6	5	5	37	-2	39
Baltimore, MD	0	0	0	0	-2	-3	-3	-3	0	3	4	5	4	8	5	4	23	-12	35
Boston, MA	0	0	0	0	-1	-1	-1	-1	0	0	1	2	1	2	1	3	6	-7	13
Cleveland, OH	0	0	0	-1	-1	-3	-4	-2	-1	2	2	2	2	2	1	1	0	-15	14
Denver, CO	0	0	0	0	0	0	0	0	0	0	0	0	1	0	0	0	1	-2	3
Detroit, MI	0	0	-1	0	-2	-5	-5	-3	1	1	2	4	2	4	11		3	-24	27
Houston, TX	0	-2	-7	-13	-17	-17	-12	-7	-1	3	6	4	5	5	6	1	-46	-76	31
Los Angeles, CA	0	0	-12	-30	-46	-37	-43	-24	-24	-8	2	11	8	6	6	7	-184	-225	41
New York, NY	0	0	0	-16	-27	-27	-37	-22	6	30	34	34	33	48	42	43	150	-148	298
Philadelphia, PA	0	0	-1	-3	-4	-12	-12	-7	-1	4	13	16	12	19	19	20	64	-46	110
Sacramento, CA	0	0	0	-2	-5	-4	-3	-1	2	4	5	6	4	3	2	1	11	-16	27
St. Louis, MO	0	0	-1	0	-1	-1	-1	0	2	3	5	5	4	3	3	6	28	-5	33

Decrease 75 to 70

Study area	Daily 8hr Max Ozone Level (ppb)																Total	Change in risk	
	0-5	5-10	10-15	15-20	20-25	25-30	30-35	35-40	40-45	45-50	50-55	55-60	60-65	65-70	70-75	>75		Inc.	Dec.
Atlanta, GA	0	0	0	0	0	0	0	1	1	2	3	2	1	0	0	0	12	0	12
Baltimore, MD	0	0	0	0	0	0	0	0	1	2	3	3	2	1	1	0	13	-1	14
Boston, MA	0	0	0	0	0	0	0	0	1	1	1	1	1	0	1	1	7	-1	8
Cleveland, OH	0	0	0	0	0	0	0	0	2	3	2	2	2	1	1	0	14	-2	15
Denver, CO	0	0	0	0	0	0	0	0	0	0	0	1	1	0	0	0	2	0	2
Detroit, MI	0	0	0	0	0	0	-1	0	2	5	3	3	4	2	3	1	23	-2	24
Houston, TX	0	0	0	0	-1	-1	-2	0	2	2	2	2	0	0	0	0	5	-4	9
Los Angeles, CA	0	0	0	0	0	0	0	0	6	16	17	2	1	0	0	0	43	0	43
New York, NY	0	0	0	0	-1	-2	0	12	27	32	35	25	5	0	0	0	134	-11	146
Philadelphia, PA	0	0	0	0	0	-1	0	0	3	7	8	9	6	3	1	1	35	-3	38
Sacramento, CA	0	0	0	0	0	-1	-1	1	2	2	2	1	0	0	0	0	7	-1	8
St. Louis, MO	0	0	0	0	0	0	0	1	2	3	4	3	2	2	1	0	18	0	19

Decrease 75 to 65

Study area	Daily 8hr Max Ozone Level (ppb)																Total	Change in risk	
	0-5	5-10	10-15	15-20	20-25	25-30	30-35	35-40	40-45	45-50	50-55	55-60	60-65	65-70	70-75	>75		Inc.	Dec.
Atlanta, GA	0	0	0	0	0	0	1	1	3	4	6	4	2	1	1	0	21	0	22
Baltimore, MD	0	0	0	0	0	0	0	0	3	4	5	7	5	1	1	0	27	-2	28
Boston, MA	0	0	0	0	0	-1	0	1	2	3	3	2	3	1	2	2	20	-2	22
Cleveland, OH	0	0	0	0	0	0	0	2	6	7	5	5	4	2	2	0	32	-3	34
Denver, CO	0	0	0	0	0	0	0	0	0	0	1	2	1	0	0	0	4	-1	5
Detroit, MI	0	0	0	0	0	0	-1	1	4	8	6	6	7	3	6	3	42	-3	45
Houston, TX	0	0	0	0	-2	-3	-3	0	4	5	4	5	0	0	0	0	11	-9	20
Los Angeles, CA	0	0	0	0	0	0	0	0	13	33	35	4	1	1	0	0	87	0	87
New York, NY	0	0	0	0	-1	2	24	85	149	136	136	90	19	0	0	0	640	-6	646
Philadelphia, PA	0	0	0	0	0	-1	0	0	7	15	17	18	12	6	1	2	76	-5	81
Sacramento, CA	0	0	0	0	0	-1	-1	2	3	4	3	2	0	0	0	0	13	-3	15
St. Louis, MO	0	0	0	0	0	0	0	3	4	7	8	6	4	4	2	1	39	-1	39

Decrease 75 to 60

Study area	Daily 8hr Max Ozone Level (ppb)																Total	Change in risk	
	0-5	5-10	10-15	15-20	20-25	25-30	30-35	35-40	40-45	45-50	50-55	55-60	60-65	65-70	70-75	>75		Inc.	Dec.
Atlanta, GA	0	0	0	0	0	0	1	2	4	7	9	6	3	1	1	0	34	0	34
Baltimore, MD	0	0	0	0	0	0	0	1	5	7	9	11	8	2	2	1	45	-2	47
Boston, MA	0	0	0	0	0	-1	1	2	4	5	5	3	5	2	3	2	32	-2	34
Cleveland, OH	0	0	0	0	-1	0	0	5	12	14	10	10	8	4	3	0	64	-3	67
Denver, CO	0	0	0	0	0	0	0	0	0	1	2	3	2	1	0	0	8	-1	8
Detroit, MI	0	0	0	0	0	0	-1	2	8	14	10	9	11	5	9	4	69	-4	73
Houston, TX	0	0	0	0	-3	-5	-4	2	8	10	7	8	1	0	0	0	24	-13	37
Los Angeles, CA	0	0	0	0	0	0	0	2	41	59	48	6	2	1	0	0	159	0	159
New York, NY								NA											
Philadelphia, PA	0	0	0	0	0	-2	-1	1	11	23	26	27	17	8	2	2	116	-7	123
Sacramento, CA	0	0	0	0	0	-2	-2	4	6	7	5	3	1	0	0	0	23	-4	27
St. Louis, MO	0	0	0	0	0	0	0	5	7	11	12	9	6	7	3	1	60	-1	61

NA: for NYC, the model-based adjustment methodology was unable to estimate ozone distributions which would meet the lower alternative standard level of 60 ppb.

Figure 7B-3 Core Short-Term Ozone-Attributable Mortality (2009) (heat map tables – absolute ozone-attributable incidence) (Smith et al., 2009)

Recent conditions

Study area	Daily 8hr Max Ozone Level (ppb)																Total
	0-5	5-10	10-15	15-20	20-25	25-30	30-35	35-40	40-45	45-50	50-55	55-60	60-65	65-70	70-75	>75	
Atlanta, GA	0	0	1	4	7	19	15	24	44	38	30	29	22	10	2	0	245
Baltimore, MD	0	0	2	2	11	23	39	40	39	56	79	65	30	14	9	0	409
Boston, MA	0	0	1	1	11	33	37	56	50	54	43	10	3	6	9	3	317
Cleveland, OH	0	0	0	6	13	27	32	51	78	56	56	39	11	24	3	0	397
Denver, CO	0	0	0	0	0	2	2	3	11	16	20	19	8	1	1	0	83
Detroit, MI	0	0	1	9	7	26	46	67	114	148	38	51	46	0	21	6	579
Houston, TX	0	1	8	20	39	79	94	99	70	65	62	48	24	17	7	8	642
Los Angeles, CA	0	1	6	19	38	66	113	84	102	182	167	125	112	68	20	18	1 122
New York, NY	0	0	4	81	144	217	281	325	407	320	323	210	134	102	21	0	2 569
Philadelphia, PA	0	0	5	13	30	75	120	91	158	149	155	168	92	41	9	0	1 106
Sacramento, CA	0	0	4	6	15	27	37	35	49	49	43	30	23	33	20	6	377
St. Louis, MO	0	0	2	7	5	19	26	57	45	69	75	48	21	11	0	0	384

Current Standard (75)

Study area	Daily 8hr Max Ozone Level (ppb)																Total
	0-5	5-10	10-15	15-20	20-25	25-30	30-35	35-40	40-45	45-50	50-55	55-60	60-65	65-70	70-75	>75	
Atlanta, GA	0	0	1	2	8	16	18	33	49	44	29	30	9	2	0	0	241
Baltimore, MD	0	0	0	1	3	13	41	71	64	92	64	45	11	0	0	0	404
Boston, MA	0	0	1	0	11	25	45	57	50	53	48	7	3	6	9	3	319
Cleveland, OH	0	0	0	0	5	25	46	68	75	81	57	28	11	6	0	0	401
Denver, CO	0	0	0	0	0	1	2	4	9	17	22	20	6	2	1	0	83
Detroit, MI	0	0	1	9	7	26	46	67	114	148	38	51	46	0	21	6	579
Houston, TX	0	0	0	6	28	51	122	123	114	90	84	36	27	7	4	4	695
Los Angeles, CA	0	0	0	0	0	0	2	17	281	328	496	152	9	0	0	0	1 285
New York, NY	0	0	0	6	36	215	427	356	632	469	274	175	56	0	0	0	2 645
Philadelphia, PA	0	0	0	2	16	51	159	126	219	175	198	97	68	0	0	0	1 112
Sacramento, CA	0	0	0	0	2	23	64	69	73	56	42	33	6	0	0	0	367
St. Louis, MO	0	0	1	6	7	17	29	54	52	77	66	53	13	8	0	0	383

Alternative Standard 70

Study area	Daily 8hr Max Ozone Level (ppb)																Total
	0-5	5-10	10-15	15-20	20-25	25-30	30-35	35-40	40-45	45-50	50-55	55-60	60-65	65-70	70-75	>75	
Atlanta, GA	0	0	0	1	9	17	21	45	57	33	29	19	1	0	0	0	233
Baltimore, MD	0	0	0	0	3	14	39	77	83	89	72	20	0	0	0	0	397
Boston, MA	0	0	0	1	2	30	39	65	60	57	43	5	0	9	9	0	320
Cleveland, OH	0	0	0	0	2	25	56	76	86	79	49	8	8	0	0	0	390
Denver, CO	0	0	0	0	0	0	2	4	10	17	30	16	3	1	0	0	83
Detroit, MI	0	0	0	0	12	13	42	73	104	175	84	64	9	19	5	0	600
Houston, TX	0	0	0	3	24	45	124	145	116	113	81	25	12	3	4	0	696
Los Angeles, CA	0	0	0	0	0	0	2	40	331	502	308	61	0	0	0	0	1 244
New York, NY	0	0	0	0	37	177	478	532	739	379	224	0	0	0	0	0	2 566
Philadelphia, PA	0	0	0	0	17	45	147	172	206	243	172	84	7	0	0	0	1 094
Sacramento, CA	0	0	0	0	1	15	76	80	79	47	50	13	0	0	0	0	361
St. Louis, MO	0	0	0	4	9	14	34	63	72	72	64	31	10	0	0	0	375

Alternative Standard 65

Study area	Daily 8hr Max Ozone Level (ppb)																Total
	0-5	5-10	10-15	15-20	20-25	25-30	30-35	35-40	40-45	45-50	50-55	55-60	60-65	65-70	70-75	>75	
Atlanta, GA	0	0	0	1	9	13	33	52	63	25	28	1	0	0	0	0	225
Baltimore, MD	0	0	0	0	1	12	43	86	108	75	57	5	0	0	0	0	387
Boston, MA	0	0	0	0	1	30	47	64	70	57	25	2	8	9	0	0	314
Cleveland, OH	0	0	0	0	1	24	80	83	91	70	16	8	0	0	0	0	373
Denver, CO	0	0	0	0	0	0	1	4	11	24	31	8	1	0	0	0	81
Detroit, MI	0	0	0	0	11	10	39	87	142	174	66	33	18	5	0	0	586
Houston, TX	0	0	0	0	11	44	137	164	137	127	48	14	6	3	0	0	692
Los Angeles, CA	0	0	0	0	0	0	2	91	402	532	161	8	0	0	0	0	1 197
New York, NY	0	0	0	0	38	462	731	943	50	0	0	0	0	0	0	0	2 224
Philadelphia, PA	0	0	0	0	15	42	138	232	261	232	115	34	0	0	0	0	1 070
Sacramento, CA	0	0	0	0	0	14	81	98	76	42	41	3	0	0	0	0	355
St. Louis, MO	0	0	0	1	12	12	41	76	86	64	56	14	0	0	0	0	362

Alternative Standard 60

Study area	Daily 8hr Max Ozone Level (ppb)																Total
	0-5	5-10	10-15	15-20	20-25	25-30	30-35	35-40	40-45	45-50	50-55	55-60	60-65	65-70	70-75	>75	
Atlanta, GA	0	0	0	0	9	14	42	56	57	30	10	0	0	0	0	0	218
Baltimore, MD	0	0	0	0	1	10	53	105	108	73	27	0	0	0	0	0	376
Boston, MA	0	0	0	0	1	32	60	65	78	45	9	7	8	0	0	0	306
Cleveland, OH	0	0	0	0	1	28	108	82	93	34	7	0	0	0	0	0	353
Denver, CO	0	0	0	0	0	0	0	7	17	43	7	0	0	0	0	0	76
Detroit, MI	0	0	0	0	5	16	39	121	164	156	46	13	4	0	0	0	565
Houston, TX	0	0	0	0	5	38	134	208	181	92	20	3	0	0	0	0	681
Los Angeles, CA	0	0	0	0	0	0	7	332	361	404	18	0	0	0	0	0	1 123
New York, NY								NA									
Philadelphia, PA	0	0	0	0	6	31	148	290	298	191	82	0	0	0	0	0	1 046
Sacramento, CA	0	0	0	0	0	9	86	119	71	55	6	0	0	0	0	0	347
St. Louis, MO	0	0	0	0	13	13	58	93	79	71	19	0	0	0	0	0	346

NA: for NYC, the model-based adjustment methodology was unable to estimate ozone distributions which would meet the lower alternative standard level of 60 ppb.

Figure 7B-4 Core Short-Term Ozone-Attributable Mortality (2009) (heat map tables – change in absolute ozone-attributable incidence) (Smith et al., 2009)

Note: negative values are risk increases, positive values are risk reductions

Decrease recent conditions to 75

Study area	Daily 8hr Max Ozone Level (ppb)																Total	Change in risk	
	0-5	5-10	10-15	15-20	20-25	25-30	30-35	35-40	40-45	45-50	50-55	55-60	60-65	65-70	70-75	>75		Inc.	Dec.
Atlanta, GA	0	0	0	-1	0	-1	0	0	1	1	2	1	1	1	0	0	4	-3	7
Baltimore, MD	0	0	-1	-1	-3	-4	-4	-1	0	3	6	5	3	2	1	0	6	-18	24
Boston, MA	0	0	0	0	0	-1	-1	-1	0	0	0	0	0	0	0	0	-2	-3	1
Cleveland, OH	0	0	0	-3	-3	-4	-3	-1	0	1	3	2	1	2	0	0	-5	-17	12
Denver, CO	0	0	0	0	0	0	0	0	0	0	0	0	0	0	0	0	0	-1	0
Detroit, MI	0	0	0	0	0	0	0	0	0	0	0	0	0	0	0	0	0	0	0
Houston, TX	0	-1	-5	-8	-10	-14	-12	-9	-3	-1	1	2	1	1	1	1	-55	-62	8
Los Angeles, CA	0	-4	-14	-25	-32	-38	-43	-22	-16	-13	2	8	12	11	4	5	-165	-208	43
New York, NY	0	0	-2	-30	-41	-37	-24	-16	-3	9	21	17	14	12	3	0	-77	-172	95
Philadelphia, PA	0	0	-3	-5	-8	-12	-9	-6	2	4	7	12	7	4	1	0	-6	-49	43
Sacramento, CA	0	0	-3	-3	-5	-5	-3	0	2	4	5	4	4	6	4	1	10	-20	31
St. Louis, MO	0	0	0	0	0	-1	0	0	0	1	1	1	0	0	0	0	1	-2	3

Decrease 75 to 70

Study area	Daily 8hr Max Ozone Level (ppb)																Total	Change in risk	
	0-5	5-10	10-15	15-20	20-25	25-30	30-35	35-40	40-45	45-50	50-55	55-60	60-65	65-70	70-75	>75		Inc.	Dec.
Atlanta, GA	0	0	0	0	0	0	0	1	2	3	2	2	1	0	0	0	8	-2	10
Baltimore, MD	0	0	0	0	0	0	-1	0	1	3	2	2	1	0	0	0	7	-2	9
Boston, MA	0	0	0	0	-1	-1	-1	0	0	1	1	0	0	0	0	0	-1	-6	5
Cleveland, OH	0	0	0	0	0	-1	0	1	2	3	3	2	1	0	0	0	12	-2	14
Denver, CO	0	0	0	0	0	0	0	0	0	0	1	0	0	0	0	0	0	-1	1
Detroit, MI	0	0	-1	-3	-2	-6	-5	-5	-5	-2	0	2	2	0	2	1	-21	-29	8
Houston, TX	0	0	0	-1	-2	-3	-3	-1	1	2	3	2	2	0	0	0	-1	-10	10
Los Angeles, CA	0	0	0	0	0	0	0	0	4	11	20	6	0	0	0	0	41	0	41
New York, NY	0	0	0	-1	-3	-14	-8	8	22	31	22	19	6	0	0	0	84	-38	121
Philadelphia, PA	0	0	0	0	-1	-2	-2	-1	5	5	8	5	4	0	0	0	19	-9	28
Sacramento, CA	0	0	0	0	0	-1	-1	1	2	2	2	2	0	0	0	0	6	-2	8
St. Louis, MO	0	0	0	-1	-1	-1	0	1	1	3	3	3	1	1	0	0	8	-4	12

Decrease 75 to 65

Study area	Daily 8hr Max Ozone Level (ppb)																Total	Change in risk	
	0-5	5-10	10-15	15-20	20-25	25-30	30-35	35-40	40-45	45-50	50-55	55-60	60-65	65-70	70-75	>75		Inc.	Dec.
Atlanta, GA	0	0	0	0	-1	-1	0	1	3	5	3	4	2	0	0	0	16	-3	19
Baltimore, MD	0	0	0	0	0	-1	-1	0	3	6	5	4	1	0	0	0	17	-3	21
Boston, MA	0	0	0	0	-1	-1	-1	0	1	2	3	1	0	0	1	0	5	-6	11
Cleveland, OH	0	0	0	0	-1	-1	0	4	5	8	7	4	2	1	0	0	29	-3	32
Denver, CO	0	0	0	0	0	0	0	0	0	1	2	1	0	0	0	0	2	-1	3
Detroit, MI	0	0	-1	-4	-2	-7	-5	-4	-2	3	2	4	5	0	3	1	-6	-27	21
Houston, TX	0	0	0	-2	-4	-5	-5	-1	2	4	6	3	3	1	1	1	4	-18	21
Los Angeles, CA	0	0	0	0	0	0	0	0	10	23	42	14	1	0	0	0	89	0	89
New York, NY	0	0	0	-1	-5	-17	16	52	107	120	81	63	21	0	0	0	437	-43	479
Philadelphia, PA	0	0	0	-1	-2	-4	-4	-1	10	11	17	10	8	0	0	0	44	-15	59
Sacramento, CA	0	0	0	0	0	-2	-1	2	3	4	3	3	1	0	0	0	12	-4	16
St. Louis, MO	0	0	-1	-2	-1	-1	0	2	3	6	6	6	2	1	0	0	21	-5	26

Decrease 75 to 60

Study area	Daily 8hr Max Ozone Level (ppb)																Total	Change in risk	
	0-5	5-10	10-15	15-20	20-25	25-30	30-35	35-40	40-45	45-50	50-55	55-60	60-65	65-70	70-75	>75		Inc.	Dec.
Atlanta, GA	0	0	-1	0	-1	-1	0	2	5	6	5	5	2	0	0	0	23	-3	26
Baltimore, MD	0	0	0	0	-1	-1	-1	1	5	10	8	6	2	0	0	0	28	-4	32
Boston, MA	0	0	0	0	-2	-1	-1	2	2	5	5	1	1	1	2	1	14	-6	20
Cleveland, OH	0	0	0	0	-1	-1	1	7	9	13	11	6	2	1	0	0	49	-3	53
Denver, CO	0	0	0	0	0	0	0	0	0	1	2	3	1	0	0	0	7	-1	8
Detroit, MI	0	0	-1	-4	-2	-7	-5	-2	2	10	4	8	8	0	4	2	15	-26	41
Houston, TX	0	0	0	-2	-7	-7	-7	0	5	8	11	6	5	1	1	1	14	-25	40
Los Angeles, CA	0	0	0	0	0	0	0	2	32	44	63	22	1	0	0	0	164	0	164
New York, NY									NA										
Philadelphia, PA	0	0	0	-1	-3	-6	-4	0	16	17	25	14	11	0	0	0	69	-20	88
Sacramento, CA	0	0	0	0	-1	-3	-1	3	6	6	5	5	1	0	0	0	21	-6	27
St. Louis, MO	0	0	-1	-2	-1	-2	0	3	5	10	9	9	3	2	0	0	37	-6	43

NA: for NYC, the model-based adjustment methodology was unable to estimate ozone distributions which would meet the lower alternative standard level of 60 ppb.

Table 7B-3 Core Short-Term Ozone-Attributable Morbidity – Hospital Admissions (2007 and 2009)

Endpoint/Study Area/Descriptor	Air Quality Scenario								
	Absolute Ozone-Attributable Incidence					Change in Ozone-Attributable Incidence			
	Base	75ppb	70ppb	65ppb	60ppb	Base-75	75-70	75-65	75-60
2007 Simulation Year									
HA (respiratory); Detroit (Katsouyanni et al., 2009)									
1hr max, penalized splines	250	230	220	210	200	18	13	23	37
1hr max, natural splines	240	230	210	200	190	17	12	22	36
HA (respiratory); NYC (Silverman and Ito, 2010; Lin et al., 2008)									
HA Chronic Lung Disease (Lin)	130	120	120	95	95	12	6.7	29	29
HA Asthma (Silverman)	450	420	400	330	330	50	28	120	120
HA Asthma, PM2.5 (Silverman)	340	310	290	240	240	36	20	84	84
HA (respiratory); LA (Linn et al., 2000)									
1hr max penalized splines	610	790	770	750	730	-180	19	38	60
HA (COPD less asthma); all 12 study areas (Medina-Ramon, et al., 2006)									
Atlanta, GA	79	67	64	61	57	12	4	6	10
Baltimore, MD	83	77	74	71	67	7	3	6	9
Boston, MA	100	100	99	95	92	3	2	6	10
Cleveland, OH	62	61	58	56	50	1	2	5	10
Denver, CO	27	27	26	25	24	1	1	2	3
Detroit, MI	91	90	87	85	81	1	3	6	9
Houston, TX	65	68	67	66	64	-2	1	2	4
Los Angeles, CA	190	180	180	170	160	3	8	16	25
New York, NY	190	180	170	130	130	19	11	50	NA
Philadelphia, PA	140	130	130	120	120	12	4	10	15
Sacramento, CA	39	34	33	32	30	5	1	2	4
St. Louis, MO	57	53	50	48	45	5	3	5	8
2009 Simulation Year									
HA (respiratory); Detroit (Katsouyanni et al., 2009)									
1hr max, penalized splines	220	220	210	200	190	0	3.6	13	25
1hr max, natural splines	210	210	210	200	190	0	3.4	12	24
HA (respiratory); NYC (Silverman and Ito, 2010; Lin et al., 2008)									
HA Chronic Lung Disease (Lin)	120	120	110	97	97	0.045	5.1	21	21
HA Asthma (Silverman)	410	410	390	340	340	8.1	24	96	96
HA Asthma, PM2.5 (Silverman)	310	300	290	250	250	5.8	17	68	68
HA (respiratory); LA (Linn et al., 2000)									
1hr max penalized splines	640	830	820	800	770	-190	18	39	62
HA (COPD less asthma); all 12 study areas (Medina-Ramon, et al., 2006)									
Atlanta, GA	65	64	61	58	56	2	3	5	8
Baltimore, MD	74	71	69	67	65	3	2	4	6
Boston, MA	92	92	92	91	89	-1	0	1	4
Cleveland, OH	58	58	56	53	50	0	2	5	8
Denver, CO	27	27	26	25	24	0	0	1	3
Detroit, MI	81	81	85	83	80	0	-3	-2	1
Houston, TX	71	74	74	73	71	-4	1	2	4
Los Angeles, CA	200	190	190	180	170	5	8	16	26
New York, NY	170	170	160	140	140	0	7	35	NA
Philadelphia, PA	120	120	120	110	110	2	2	6	9
Sacramento, CA	41	36	35	34	32	5	1	3	4
St. Louis, MO	51	50	49	47	44	0	2	4	6

NA: for NYC, the model-based adjustment methodology was unable to estimate ozone distributions which would meet the lower alternative standard level of 60 ppb.

Table 7B-4 Core Short-Term Ozone-Attributable Morbidity – Emergency Room Visits (2007 and 2009)

Endpoint/Study Area/Descriptor	Air Quality Scenario								
	Absolute Ozone-Attributable Incidence					Change in Ozone-Attributable Incidence			
	Base	75ppb	70ppb	65ppb	60ppb	Base-75	75-70	75-65	75-60
2007 Simulation Year									
ER Visits (repiratory); Atlanta (Strickland et al., 2007)									
Distributed lag 0-7 days	8,500	7,500	7,200	6,900	6,600	1,300	410	740	1,200
Average day lag 0-2	5,100	4,500	4,200	4,100	3,900	750	230	420	670
ER-visits (respiratory); Atlanta (Tolbert et al., 2007, Darrow et al., 2011)									
Tolbert	9,200	8,100	7,800	7,500	7,100	1,100	360	670	1,100
Tolbert-CO	8,200	7,200	6,900	6,700	6,300	1,000	320	590	940
Tolbert-NO2	7,400	6,500	6,300	6,000	5,700	920	290	530	840
Tolbert-PM10	5,800	5,100	4,900	4,700	4,500	720	230	420	660
Tolbert-PM10, NO2	5,600	4,900	4,700	4,600	4,300	690	220	400	640
Darrow	5,000	4,400	4,200	4,000	3,800	610	190	360	560
ER-visits (asthma); NYC (Ito et al, 2007)									
single pollutant model	9,600	9,000	8,600	7,100	NA	780	530	2,300	NA
PM2.5	7,600	7,100	6,700	5,500		610	410	1,800	
NO2	6,300	5,800	5,600	4,500		500	330	1,500	
CO	10,000	9,500	9,100	7,500		830	570	2,500	
SO2	7,800	7,300	6,900	5,700		630	420	1,900	
2009 Simulation Year									
ER Visits (repiratory); Atlanta (Strickland et al., 2007)									
Distributed lag 0-7 days	6,900	6,800	6,600	6,400	6,200	170	310	570	800
Average day lag 0-2	4,100	4,000	3,900	3,700	3,600	98	170	320	460
ER-visits (respiratory); Atlanta (Tolbert et al., 2007, Darrow et al., 2011)									
Tolbert (single pollutant	7,500	7,400	7,200	6,900	6,700	140	270	500	720
Tolbert-CO	6,700	6,600	6,400	6,200	6,000	120	240	450	640
Tolbert-NO2	6,100	6,000	5,800	5,600	5,400	110	210	400	580
Tolbert-PM10	4,800	4,700	4,500	4,400	4,200	88	170	310	450
Tolbert-PM10, NO2	4,600	4,500	4,400	4,200	4,100	85	160	300	430
Darrow (single pollutant	4,100	4,000	3,900	3,700	3,600	75	140	270	380
ER-visits (asthma); NYC (Ito et al, 2007)									
single pollutant model	8,700	8,800	8,500	7,300	NA	-72	400	1,800	NA
PM2.5	6,800	6,900	6,600	5,700		-53	310	1,400	
NO2	5,600	5,700	5,500	4,700		-42	250	1,100	
CO	9,200	9,300	9,000	7,700		-77	430	1,900	
SO2	7,000	7,100	6,800	5,900		-55	320	1,400	

NA: for NYC, the model-based adjustment methodology was unable to estimate ozone distributions which would meet the lower alternative standard level of 60 ppb.

Table 7B-5 Core Short-Term Ozone-Attributable Morbidity – Asthma Exacerbations (2007 and 2009)

Endpoint/Study Area/Descriptor	Air Quality Scenario								
	Absolute Ozone-Attributable Incidence					Change in Ozone-Attributable Incidence			
	Base	75ppb	70ppb	65ppb	60ppb	Base-75	75-70	75-65	75-60
2007 Simulation Year									
Asthma exacerbation (wheeze); Boston (Gent et al., 2003, 2004)									
Chest Tightness	70,000	69,000	67,000	65,000	63,000	2,500	2,100	5,600	8,600
Shortness of Breath	50,000	49,000	48,000	46,000	44,000	1,600	1,400	3,700	5,700
Chest Tightness (1hr max)	51,000	51,000	50,000	48,000	47,000	900	1,200	3,200	5,000
Shortness of Breath (1hr max)	59,000	59,000	58,000	56,000	54,000	1,000	1,300	3,600	5,800
Chest Tightness (PM2.5)	71,000	69,000	68,000	66,000	63,000	2,500	2,100	5,600	8,700
Chest Tightness (PM2.5)	66,000	64,000	63,000	60,000	58,000	2,300	1,900	5,100	8,000
Wheeze (PM2.5)	130,000	130,000	130,000	120,000	120,000	4,500	3,800	10,000	16,000
2009 Simulation Year									
Asthma exacerbation (wheeze); Boston (Gent et al., 2003, 2004)									
Chest Tightness	63,000	63,000	63,000	62,000	60,000	-150	490	2,400	4,800
Shortness of Breath	45,000	45,000	44,000	43,000	42,000	-100	330	1,600	3,200
Chest Tightness (1hr max)	46,000	47,000	47,000	46,000	45,000	-360	-180	790	2,200
Shortness of Breath (1hr max)	54,000	54,000	54,000	53,000	52,000	-420	-210	910	2,500
Chest Tightness (PM2.5)	64,000	64,000	64,000	63,000	61,000	-160	500	2,400	4,800
Chest Tightness (PM2.5)	59,000	59,000	59,000	58,000	56,000	-140	450	2,200	4,400
Wheeze (PM2.5)	120,000	120,000	120,000	120,000	110,000	-280	900	4,300	8,700

Table 7B-6 Core Long-Term Ozone-Attributable Respiratory Mortality (2007) (incidence, percent of baseline mortality, incidence per 100,000) (Jerrett et al., 2009)

Study Area	Air Quality Scenario								
	Absolute Ozone-Attributable Incidence					Change in Ozone-Attributable Incidence			
	Base	75ppb	70ppb	65ppb	60ppb	Base-75	75-70	75-65	75-60
Atlanta, GA	840	710	680	650	610	150	43	78	120
	(300 - 1300)	(260 - 1100)	(240 - 1100)	(230 - 1000)	(220 - 960)	(51 - 250)	(15 - 71)	(26 - 130)	(41 - 200)
Baltimore, MD	820	750	720	700	660	78	33	67	110
	(300 - 1300)	(270 - 1200)	(260 - 1100)	(250 - 1100)	(240 - 1000)	(27 - 130)	(11 - 55)	(23 - 110)	(37 - 180)
Boston, MA	1,200	1,100	1,100	1,000	1,000	41	35	93	140
	(410 - 1800)	(400 - 1800)	(390 - 1700)	(370 - 1600)	(350 - 1600)	(14 - 69)	(12 - 58)	(32 - 150)	(49 - 240)
Cleveland, OH	540	530	500	480	440	21	26	56	100
	(190 - 850)	(190 - 820)	(180 - 790)	(170 - 750)	(160 - 690)	(7 - 34)	(9 - 43)	(19 - 93)	(35 - 170)
Denver, CO	500	480	470	450	430	23	19	39	64
	(180 - 770)	(170 - 740)	(170 - 720)	(160 - 700)	(150 - 670)	(8 - 39)	(6 - 31)	(13 - 64)	(22 - 100)
Detroit, MI	790	760	730	710	680	30	35	63	99
	(280 - 1200)	(270 - 1200)	(260 - 1100)	(250 - 1100)	(240 - 1100)	(10 - 50)	(12 - 59)	(21 - 100)	(33 - 160)
Houston, TX	560	550	540	530	520	22	9.5	19	32
	(200 - 880)	(190 - 860)	(190 - 850)	(190 - 830)	(180 - 820)	(7 - 36)	(3 - 16)	(6 - 31)	(11 - 53)
Los Angeles, CA	2,600	2,600	2,500	2,400	2,200	95	140	260	410
	(960 - 4100)	(930 - 4000)	(890 - 3800)	(850 - 3700)	(800 - 3500)	(32 - 160)	(46 - 230)	(89 - 430)	(140 - 670)
New York, NY	2,100	1,800	1,700	1,400	1,400	280	120	480	NA
	(750 - 3200)	(660 - 2900)	(620 - 2700)	(500 - 2200)	(500 - 2200)	(96 - 470)	(41 - 200)	(160 - 790)	NA
Philadelphia, PA	1,400	1,300	1,200	1,200	1,100	170	56	120	170
	(500 - 2100)	(450 - 1900)	(430 - 1900)	(410 - 1800)	(390 - 1700)	(57 - 270)	(19 - 93)	(40 - 190)	(59 - 290)
Sacramento, CA	790	680	660	630	600	130	31	60	100
	(290 - 1200)	(250 - 1100)	(240 - 1000)	(230 - 1000)	(210 - 940)	(44 - 210)	(10 - 51)	(20 - 98)	(34 - 170)
St. Louis, MO	640	600	570	540	510	60	34	69	100
	(230 - 1000)	(210 - 930)	(200 - 890)	(190 - 840)	(180 - 800)	(20 - 99)	(230 - 1000)	(23 - 110)	(35 - 170)

Study Area	Air Quality Scenario								
	Percent of Baseline Incidence Attributable to Ozone					Change in O_3-Attributable Risk			
	Base	75ppb	70ppb	65ppb	60ppb	Base-75	75-70	75-65	75-60
Atlanta, GA	21.7	18.6	17.7	16.9	15.9	14	5	9	15
Baltimore, MD	20 3	18.7	18.1	17.4	16.5	8	4	8	12
Boston, MA	17.7	17 2	16.7	16.0	15.3	3	3	7	11
Cleveland, OH	18 2	17.6	16.9	16.1	14.7	3	4	9	17
Denver, CO	21 5	20.8	20.1	19.4	18.5	4	3	7	11
Detroit, MI	19 0	18.4	17.7	17.1	16.4	3	4	7	11
Houston, TX	16 8	16.3	16.1	15.8	15.5	3	1	3	5
Los Angeles, CA	21 0	20.4	19.6	18.7	17.8	3	4	9	13
New York, NY	19 0	16.9	15.9	13.0	13.0	11	6	24	NA
Philadelphia, PA	20 3	18.4	17.7	16.9	16.2	10	4	8	12
Sacramento, CA	20.4	17.7	17.1	16.5	15.6	13	4	7	13
St. Louis, MO	20 2	18.7	17.9	16.9	16.0	7	5	10	15

Study Area	Air Quality Scenario								
	Ozone-Attributable Deaths per 100,000					Change in Ozone-Attributable Deaths per 100,000			
	Base	75ppb	70ppb	65ppb	60ppb	Base-75	75-70	75-65	75-60
Atlanta, GA	25.47	21.75	20.67	19.78	18.66	4.58	1.31	2 38	3.69
Baltimore, MD	25.52	23.53	22.68	21.81	20.68	2.46	1.04	2 09	3.42
Boston, MA	25.33	24.58	23.94	22.86	21 90	0.91	0.77	2 05	3.17
Cleveland, OH	25.77	24.96	23.94	22.72	20 83	0.98	1.23	2.67	4.85
Denver, CO	24.29	23.39	22.66	21.86	20 87	1.14	0.91	1 90	3.10
Detroit, MI	23.24	22.51	21.65	20.97	20 08	0.90	1 05	1 86	2.92
Houston, TX	16.80	16.26	16.02	15.80	15.46	0.64	0 28	0 55	0.94
Los Angeles, CA	22.58	21.94	21.01	20.12	19.11	0.81	1.16	2 25	3.46
New York, NY	21.41	18.98	17.92	14.65	14.65	2.92	1 26	4 98	NA
Philadelphia, PA	29.87	26.95	25.96	24.85	23 81	3.59	1 21	2 54	3.76
Sacramento, CA	29.29	25.39	24.44	23.55	22 23	4.75	1.15	2.21	3.75
St. Louis, MO	32.23	29.80	28.41	26.94	25.43	3.01	1.69	3.45	5.21

NA: for NYC, the model-based adjustment methodology was unable to estimate ozone distributions which would meet the lower alternative standard level of 60 ppb.

Table 7B-7 Core Long-Term Ozone-Attributable Respiratory Mortality (2009) (incidence, percent of baseline mortality, incidence per 100,000) (Jerrett et al., 2009)

Study Area	Air Quality Scenario								
	Absolute Ozone-Attributable Incidence					Change in Ozone-Attributable Incidence			
	Base	75ppb	70ppb	65ppb	60ppb	Base-75	75-70	75-65	75-60
Atlanta, GA	730	700	670	640	610	33	41	76	100
	(260 - 1100)	(250 - 1100)	(240 - 1000)	(230 - 1000)	(220 - 970)	(11 - 55)	(14 - 68)	(26 - 120)	(36 - 170)
Baltimore, MD	770	730	710	680	660	50	25	54	84
	(280 - 1200)	(260 - 1100)	(250 - 1100)	(240 - 1100)	(230 - 1000)	(17 - 83)	(8 - 41)	(18 - 90)	(28 - 140)
Boston, MA	1,100	1,100	1,100	1,000	1,000	-2.1	6.8	42	85
	(380 - 1700)	(380 - 1700)	(380 - 1700)	(370 - 1600)	(350 - 1600)	(-1 - -4)	(2 - 11)	(14 - 69)	(29 - 140)
Cleveland, OH	520	510	490	470	440	16	24	54	84
	(190 - 820)	(180 - 800)	(170 - 770)	(170 - 730)	(160 - 690)	(5 - 26)	(8 - 41)	(18 - 89)	(29 - 140)
Denver, CO	500	490	490	470	440	0.77	9.0	28	70
	(180 - 770)	(180 - 770)	(180 - 750)	(170 - 730)	(160 - 680)	(0 - 1)	(3 - 15)	(10 - 47)	(24 - 120)
Detroit, MI	720	720	730	710	680	NA	-8 9	18	51
	(260 - 1100)	(260 - 1100)	(260 - 1100)	(250 - 1100)	(240 - 1100)	NA	(-3 - -15)	(6 - 30)	(17 - 85)
Houston, TX	600	610	600	580	560	-4 9	14	30	49
	(220 - 940)	(220 - 950)	(210 - 930)	(210 - 910)	(200 - 890)	(-2 - -8)	(5 - 23)	(10 - 49)	(17 - 82)
Los Angeles, CA	2,800	2,800	2,600	2,500	2,400	110	130	280	430
	(1000 - 4400)	(1000 - 4300)	(960 - 4100)	(910 - 3900)	(860 - 3700)	(37 - 180)	(45 - 220)	(94 - 460)	(150 - 710)
New York, NY	1,900	1,900	1,800	1,500	1,500	57	110	390	NA
	(690 - 3000)	(670 - 2900)	(630 - 2800)	(540 - 2400)	(540 - 2400)	(19 - 94)	(37 - 180)	(130 - 630)	NA
Philadelphia, PA	1,300	1,200	1,200	1,100	1,100	56	44	94	140
	(450 - 2000)	(430 - 1900)	(420 - 1800)	(400 - 1800)	(390 - 1700)	(19 - 93)	(15 - 73)	(32 - 160)	(47 - 230)
Sacramento, CA	850	730	700	680	640	150	34	66	110
	(310 - 1300)	(260 - 1100)	(250 - 1100)	(240 - 1100)	(230 - 1000)	(50 - 240)	(12 - 57)	(22 - 110)	(36 - 170)
St. Louis, MO	580	580	560	530	510	7.3	24	53	86
	(210 - 910)	(210 - 900)	(200 - 870)	(190 - 840)	(180 - 800)	(3 - 12)	(210 - 910)	(18 - 87)	(29 - 140)

Study Area	Air Quality Scenario								
	Percent of Baseline Incidence Attributable to Ozone					Change in O_3-Attributable Risk			
	Base	75ppb	70ppb	65ppb	60ppb	Base-75	75-70	75-65	75-60
Atlanta, GA	17.6	17.0	16.1	15.4	14.8	4	5	9	13
Baltimore, MD	18.4	17.4	16.9	16.3	15.7	5	3	6	10
Boston, MA	15.9	15.9	15.9	15.4	14.9	0	1	3	7
Cleveland, OH	17.2	16.8	16.1	15.3	14.4	2	4	9	15
Denver, CO	20.0	20.0	19.7	19.1	17.7	0	1	5	12
Detroit, MI	17.0	17.0	17.1	16.6	16.0	NA	-1	2	6
Houston, TX	16.8	16.9	16.6	16.2	15.8	-1	2	4	7
Los Angeles, CA	21.4	20.7	19.9	19.0	18.1	3	4	8	13
New York, NY	17.1	16.7	15.9	13.7	13.7	2	5	18	NA
Philadelphia, PA	17.8	17.2	16.7	16.1	15.5	4	3	7	10
Sacramento, CA	20.9	18.0	17.3	16.6	15.8	14	4	8	12
St. Louis, MO	17.9	17.7	17.1	16.4	15.5	1	4	8	13

Study Area	Air Quality Scenario								
	Ozone-Attributable Deaths per 100,000					Change in Ozone-Attributable Deaths per 100,000			
	Base	75ppb	70ppb	65ppb	60ppb	Base-75	75-70	75-65	75-60
Atlanta, GA	21 33	20.53	19.53	18.66	17.92	0.97	1.20	2.21	3.07
Baltimore, MD	23.68	22.40	21.77	21.01	20.23	1.55	0.76	1.67	2.58
Boston, MA	23 02	23.06	22.93	22.30	21.49	-0.04	0.15	0.90	1.84
Cleveland, OH	24 89	24.26	23.29	22.10	20.85	0.76	1.16	2.56	4.00
Denver, CO	23.17	23.14	22.81	22.08	20.44	0.04	0.42	1.32	3.30
Detroit, MI	21 39	21.39	21.61	20.95	20.12	NA	-0.26	0.54	1.52
Houston, TX	17 07	17.18	16.86	16.48	16.01	-0.14	0.39	0.84	1.40
Los Angeles, CA	23.62	22.89	22.00	21.03	19.97	0.92	1.11	2.31	3.58
New York, NY	19 57	19.09	18.15	15.71	15.71	0.58	1.11	3.92	NA
Philadelphia, PA	26 50	25.51	24.74	23.84	23.05	1.19	0.94	1.99	2.93
Sacramento, CA	30.74	26.44	25.42	24.46	23.24	5.26	1.24	2.37	3.81
St. Louis, MO	28.78	28.49	27.49	26.32	24.91	0.36	1.20	2.60	4.24

NA: for NYC, the model-based adjustment methodology was unable to estimate ozone distributions which would meet the lower alternative standard level of 60 ppb.

Appendix 7-C – Detailed Sensitivity Analysis Results

Table 7C-1 Sensitivity Analysis – *ST Mortality: Smaller Smith et al., 2009-based study area (2009)* (incidence, percent of baseline mortality, incidence per 100,000) - compare with Core Results in Table 7B-2

Study Area	Air Quality Scenario								
	Absolute Ozone-Attributable Incidence					Change in Ozone-Attributable Incidence			
	Base	75ppb	70ppb	65ppb	60ppb	Base-75	75-70	75-65	75-60
Atlanta, GA	170	170	170	160	160	-2	2	6	11
	(-240 - 560)	(-240 - 570)	(-230 - 560)	(-230 - 550)	(-220 - 530)	(2 - -6)	(-3 - 7)	(-9 - 21)	(-15 - 35)
Baltimore, MD	190	200	200	200	190	-11	2	5	8
	(-100 - 470)	(-110 - 500)	(-110 - 500)	(-110 - 490)	(-110 - 480)	(6 - -29)	(-1 - 4)	(-2 - 11)	(-4 - 20)
Boston, MA	160	160	170	170	160	-4	-7	-6	-4
	(-220 - 520)	(-230 - 530)	(-240 - 560)	(-240 - 550)	(-230 - 550)	(5 - -13)	(10 - -25)	(9 - -22)	(6 - -14)
Cleveland, OH	340	360	350	340	320	-17	8	20	37
	(-33 - 710)	(-34 - 740)	(-34 - 730)	(-32 - 700)	(-31 - 670)	(2 - -35)	(-1 - 16)	(-2 - 42)	(-4 - 77)
Denver, CO	73	74	75	74	70	-1	-1	0	4
	(-240 - 370)	(-240 - 380)	(-240 - 380)	(-240 - 380)	(-230 - 360)	(3 - -4)	(3 - -5)	(0 - -1)	(-12 - 19)
Detroit, MI	530	530	550	540	530	NA	-28	-17	0
	(26 - 1000)	(26 - 1000)	(27 - 1100)	(26 - 1000)	(26 - 1000)	NA	(-1 - -54)	(-1 - -33)	(0 - 1)
Houston, TX	510	570	570	570	560	-54	-3	-1	5
	(96 - 920)	(110 - 1000)	(110 - 1000)	(110 - 1000)	(100 - 1000)	(-10 - -98)	(-1 - -5)	(0 - -2)	(1 - 9)
Los Angeles, CA	850	990	960	920	860	-140	34	73	130
	(-360 - 2000)	(-420 - 2400)	(-400 - 2300)	(-380 - 2200)	(-360 - 2100)	(57 - -330)	(-14 - 82)	(-30 - 180)	(-55 - 320)
New York, NY	1800	2100	2100	1900	1900	-260	7	210	NA
	(1100 - 2600)	(1300 - 2900)	(1300 - 2900)	(1100 - 2600)	(1100 - 2600)	(-160 - -370)	(4 - 10)	(130 - 300)	NA
Philadelphia, PA	720	770	770	760	750	-53	2	9	18
	(160 - 1300)	(170 - 1400)	(170 - 1400)	(170 - 1300)	(170 - 1300)	(-11 - -94)	(0 - 4)	(2 - 16)	(4 - 32)
Sacramento, CA	370	360	360	350	340	3	6	12	21
	(-390 - 1100)	(-390 - 1100)	(-380 - 1100)	(-370 - 1100)	(-360 - 1000)	(-3 - 8)	(-6 - 18)	(-12 - 36)	(-22 - 63)
St. Louis, MO	250	250	270	270	270	-6	-15	-18	-15
	(-62 - 550)	(-64 - 560)	(-67 - 590)	(-68 - 600)	(-67 - 590)	(1 - -13)	(4 - -33)	(5 - -41)	(4 - -34)

Study Area	Air Quality Scenario								
	Percent of Baseline Incidence Attributable to Ozone					Change in O_3-Attributable Risk			
	Base	75ppb	70ppb	65ppb	60ppb	Base-75	75-70	75-65	75-60
Atlanta, GA	1.0	1.0	1.0	1.0	1.0	-1	1	4	6
Baltimore, MD	1.7	1.7	1.7	1.7	1.7	-6	1	2	4
Boston, MA	1.1	1.1	1.1	1.1	1.1	-2	-4	-4	-2
Cleveland, OH	2.2	2.3	2.3	2.2	2.1	-5	2	5	10
Denver, CO	0.8	0.8	0.8	0.8	0.8	-1	-1	0	5
Detroit, MI	2.6	2.6	2.8	2.7	2.6	NA	-5	-3	0
Houston, TX	1.8	1.9	1.9	1.9	1.9	-10	-1	0	1
Los Angeles, CA	0.9	1.1	1.0	1.0	0.9	-16	3	7	13
New York, NY	3.7	4.2	4.1	3.8	3.8	-14	0	10	NA
Philadelphia, PA	2.6	2.8	2.8	2.8	2.8	-7	0	1	2
Sacramento, CA	1.2	1.2	1.2	1.2	1.2	1	2	3	6
St. Louis, MO	2.1	2.2	2.3	2.3	2.3	-2	-6	-7	-6

Study Area	Air Quality Scenario								
	Ozone-Attributable Deaths per 100,000					Change in Ozone-Attributable Deaths per 100,000			
	Base	75ppb	70ppb	65ppb	60ppb	Base-75	75-70	75-65	75-60
Atlanta, GA	3.81	3.81	3.81	3.59	3.59	-0.04	0.05	0.14	0.25
Baltimore, MD	9.73	10.24	10.24	10.24	9.73	-0.56	0.08	0.23	0.41
Boston, MA	4.43	4.43	4.71	4.71	4.43	-0.11	-0.20	-0.18	-0.11
Cleveland, OH	12.20	12.92	12.56	12.20	11.48	-0.61	0.27	0.72	1.33
Denver, CO	2.67	2.70	2.74	2.70	2.56	-0.03	-0.04	0.00	0.13
Detroit, MI	12.11	12.11	12.56	12.34	12.11	NA	-0.64	-0.39	0.01
Houston, TX	8.84	9.88	9.88	9.88	9.71	-0.94	-0.05	-0.02	0.08
Los Angeles, CA	5.18	6.04	5.85	5.61	5.24	-0.85	0.21	0.45	0.79
New York, NY	13.60	15.86	15.86	14.35	14.35	-1.96	0.05	1.59	NA
Philadelphia, PA	15.20	16.25	16.25	16.04	15.83	-1.12	0.04	0.19	0.38
Sacramento, CA	8.52	8.29	8.29	8.06	7.83	0.06	0.14	0.28	0.48
St. Louis, MO	13.48	13.48	14.55	14.55	14.55	-0.32	-0.81	-0.97	-0.81

NA: for NYC, the model-based adjustment methodology was unable to estimate ozone distributions which would meet the lower alternative standard level of 60 ppb.

Table 7C-1 Sensitivity Analysis – *ST Mortality: Smaller Smith et al., 2009-based study area (2009)* (heat maps for just meeting existing standard and risk reductions from just meeting alternative standards) (see Key at bottom of figure)

Current Standard (75)

Study area	Daily 8hr Max Ozone Level (ppb)																Total
	0-5	5-10	10-15	15-20	20-25	25-30	30-35	35-40	40-45	45-50	50-55	55-60	60-65	65-70	70-75	>75	
Atlanta, GA	0	0	0	2	5	11	15	18	21	29	22	19	14	7	3	4	170
Baltimore, MD	0	0	0	0	2	8	25	31	35	40	32	18	7	2	0	0	200
Boston, MA	0	0	0	0	4	13	24	28	27	29	15	6	5	3	2	2	160
Cleveland, OH	0	0	0	0	5	24	38	65	58	72	44	36	15	3	0	0	360
Denver, CO	0	0	0	0	0	1	2	5	9	13	19	16	6	2	0	0	74
Detroit, MI	0	0	2	10	13	17	49	68	106	116	51	39	26	19	5	6	526
Houston, TX	0	0	1	6	22	37	93	104	95	69	68	34	20	11	3	6	567
Los Angeles, CA	0	0	0	0	0	0	2	9	189	224	378	168	20	0	0	0	989
New York, NY	0	0	0	9	22	170	231	280	318	450	267	186	86	77	0	0	2,096
Philadelphia, PA	0	0	0	3	20	70	87	137	106	123	122	71	22	0	6	0	769
Sacramento, CA	0	0	0	0	4	21	58	73	59	57	31	36	16	6	2	0	364
St. Louis, MO	0	0	2	3	7	13	25	31	37	54	35	28	11	2	4	3	254

Decrease 75 to 70

Study area	Daily 8hr Max Ozone Level (ppb)																Total	Change in risk	
	0-5	5-10	10-15	15-20	20-25	25-30	30-35	35-40	40-45	45-50	50-55	55-60	60-65	65-70	70-75	>75		Inc.	Dec.
Atlanta, GA	0	0	0	0	0	-1	-1	0	0	1	1	1	1	0	0	0	2	-3	5
Baltimore, MD	0	0	0	0	0	0	-1	0	0	1	1	1	0	0	0	0	2	-2	4
Boston, MA	0	0	0	0	-1	-1	-1	-2	-1	-1	0	0	0	0	0	0	-7	-8	0
Cleveland, OH	0	0	0	0	0	-1	-1	0	1	3	2	2	1	0	0	0	7	-3	10
Denver, CO	0	0	0	0	0	0	0	0	0	0	0	0	0	0	0	0	-1	-2	1
Detroit, MI	0	0	-1	-3	-4	-3	-6	-6	-6	-3	0	1	1	1	0	0	-28	-32	4
Houston, TX	0	0	0	-1	-2	-2	-3	-1	0	1	2	1	1	1	0	0	-3	-10	7
Los Angeles, CA	0	0	0	0	0	0	0	0	3	7	16	7	1	0	0	0	34	0	34
New York, NY	0	0	0	-2	-3	-20	-14	-4	-1	10	13	14	7	8	0	0	7	-59	66
Philadelphia, PA	0	0	0	-1	-2	-4	-2	-1	0	3	3	3	1	0	0	0	2	-12	14
Sacramento, CA	0	0	0	0	0	-1	-1	1	1	2	1	2	1	0	0	0	6	-3	9
St. Louis, MO	0	0	-1	-1	-1	-2	-3	-2	-2	-2	-1	0	0	0	0	0	-15	-15	0

Decrease 75 to 65

Study area	Daily 8hr Max Ozone Level (ppb)																Total	Change in risk	
	0-5	5-10	10-15	15-20	20-25	25-30	30-35	35-40	40-45	45-50	50-55	55-60	60-65	65-70	70-75	>75		Inc.	Dec.
Atlanta, GA	0	0	0	-1	-1	-1	-1	0	1	2	2	2	2	1	0	1	6	-4	10
Baltimore, MD	0	0	0	0	0	-1	-1	-1	1	2	2	1	1	0	0	0	4	-4	8
Boston, MA	0	0	0	0	-1	-2	-2	-2	-1	-1	0	0	0	0	0	0	-6	-8	2
Cleveland, OH	0	0	0	0	-1	-2	-1	2	3	6	5	5	2	0	0	0	20	-4	24
Denver, CO	0	0	0	0	0	0	0	0	0	0	1	0	0	0	0	0	0	-2	2
Detroit, MI	0	0	-1	-4	-5	-4	-7	-5	-4	1	3	3	3	2	1	1	-17	-32	15
Houston, TX	0	0	0	-2	-3	-4	-6	-2	1	3	4	3	2	1	0	1	-1	-18	16
Los Angeles, CA	0	0	0	0	0	0	0	0	6	16	33	16	2	0	0	0	73	0	73
New York, NY	0	0	0	-4	-5	-38	-15	12	27	66	61	54	27	28	0	0	214	-87	300
Philadelphia, PA	0	0	0	-1	-4	-8	-4	-1	3	7	8	6	2	0	1	0	9	-21	30
Sacramento, CA	0	0	0	0	-1	-2	-1	2	3	4	2	3	2	1	0	0	12	-5	17
St. Louis, MO	0	0	-2	-1	-2	-3	-4	-3	-3	-1	0	1	0	0	0	0	-18	-20	2

Decrease 75 to 60

Study area	Daily 8hr Max Ozone Level (ppb)																Total	Change in risk	
	0-5	5-10	10-15	15-20	20-25	25-30	30-35	35-40	40-45	45-50	50-55	55-60	60-65	65-70	70-75	>75		Inc.	Dec.
Atlanta, GA	0	0	0	-1	-1	-1	-1	0	1	3	3	3	2	1	1	1	11	-4	15
Baltimore, MD	0	0	0	0	-1	-1	-2	-1	2	3	3	2	1	0	0	0	8	-5	13
Boston, MA	0	0	0	0	-1	-2	-2	-2	0	0	1	1	1	0	0	0	-4	-8	4
Cleveland, OH	0	0	0	0	-1	-2	0	4	5	11	8	8	4	1	0	0	37	-4	41
Denver, CO	0	0	0	0	0	0	0	0	0	0	1	2	1	0	0	0	4	-2	6
Detroit, MI	0	0	-2	-5	-6	-4	-6	-4	-1	6	5	5	5	3	1	1	0	-32	33
Houston, TX	0	0	0	-2	-5	-6	-8	-2	3	5	8	5	4	2	1	2	5	-26	31
Los Angeles, CA	0	0	0	0	0	0	0	1	22	32	49	25	3	0	0	0	133	0	133
New York, NY								NA											
Philadelphia, PA	0	0	0	-2	-5	-11	-5	0	5	10	12	10	4	0	1	0	18	-28	46
Sacramento, CA	0	0	0	0	-1	-3	-2	3	4	6	4	6	3	1	0	0	21	-7	28
St. Louis, MO	0	-1	-2	-2	-3	-4	-5	-3	-2	0	1	2	1	0	1	1	-15	-22	6

NA: for NYC, the model-based adjustment methodology was unable to estimate ozone distributions which would meet the lower alternative standard level of 60 ppb.

Key: For *current standard (75)* which is an absolute risk metric, color gradient ranges from blue (smallest ozone-related mortality count) to red (highest ozone-related mortality count). For *Decrease results*, color gradient ranges from red (increase in risk – negative cell values) to blue (reduction in risk – positive cell values).

Table 7C-2 Sensitivity Analysis – *ST Mortality: Alternate method for simulating standards* (2009) (incidence, percent of baseline mortality, incidence per 100,000) - compare with Core Results in Table 7B-2

Study Area	Air Quality Scenario								
	Absolute Ozone-Attributable Incidence					Change in Ozone-Attributable Incidence			
	Base	75ppb	70ppb	65ppb	60ppb	Base-75	75-70	75-65	75-60
Denver, CO	83	84	84	82	77	-1	0	2	7
	(-270 - 420)	(-280 - 430)	(-280 - 430)	(-270 - 420)	(-250 - 390)	(5 - -8)	(1 - -1)	(-5 - 8)	(-23 - 37)
Detroit, MI	580	580	590	580	550	0	-12	5	30
	(28 - 1100)	(28 - 1100)	(29 - 1100)	(28 - 1100)	(27 - 1100)	0	(-1 - -23)	(0 - 9)	(1 - 58)
Houston, TX	640	680	680	680	660	-37	-2	2	20
	(120 - 1200)	(130 - 1200)	(130 - 1200)	(130 - 1200)	(120 - 1200)	(-7 - -67)	(0 - -3)	(0 - 3)	(4 - 37)
Los Angeles, CA	1100	1200	1100	1100	1100	-31	25	53	80
	(-470 - 2700)	(-480 - 2800)	(-470 - 2700)	(-460 - 2600)	(-450 - 2600)	(13 - -76)	(-10 - 60)	(-22 - 130)	(-33 - 190)
New York, NY	2600	2600	2600	2300	NA	-2	21	250	NA
	(1500 - 3600)	(1500 - 3600)	(1500 - 3500)	(1400 - 3200)	NA	(-1 - -3)	(12 - 29)	(150 - 350)	NA
Philadelphia, PA	1100	1100	1100	1000	1000	12	21	47	79
	(240 - 1900)	(240 - 1900)	(240 - 1900)	(230 - 1800)	(220 - 1800)	(3 - 22)	(5 - 38)	(10 - 83)	(17 - 140)
Sacramento, CA	380	360	360	350	340	15	7	14	22
	(-400 - 1100)	(-390 - 1100)	(-380 - 1100)	(-370 - 1000)	(-360 - 1000)	(-16 - 44)	(-8 - 22)	(-14 - 42)	(-24 - 68)

Study Area	Air Quality Scenario								
	Percent of Baseline Incidence Attributable to Ozone					Change in O_3-Attributable Risk			
	Base	75ppb	70ppb	65ppb	60ppb	Base-75	75-70	75-65	75-60
Denver, CO	0.8	0.8	0.8	0.8	0.7	-2	0	2	8
Detroit, MI	2.70	2.70	2.75	2.68	2.56	0.00	-0.05	0.02	0.14
Houston, TX	1.77	1.87	1.88	1.87	1.82	-0.10	0.00	0.01	0.06
Los Angeles, CA	0.9	0.9	0.9	0.9	0.9	-3	2	5	7
New York, NY	3.8	3.8	3.8	3.5	NA	0	1	9	NA
Philadelphia, PA	2.95	2.91	2.86	2.80	2.71	0.03	0.06	0.12	0.21
Sacramento, CA	1.23	1.19	1.16	1.14	1.11	0.05	0.02	0.05	0.07

Study Area	Air Quality Scenario								
	Ozone-Attributable Deaths per 100,000					Change in Ozone-Attributable Deaths per 100,000			
	Base	75ppb	70ppb	65ppb	60ppb	Base-75	75-70	75-65	75-60
Denver, CO	2.23	2.26	2.26	2.21	2.07	-0.04	0.00	0.04	0.19
Detroit, MI	10.32	10.32	10.50	10.32	9.78	NA	-0.21	0.08	0.53
Houston, TX	9.94	10.56	10.56	10.56	10.25	-0.57	-0.03	0.03	0.31
Los Angeles, CA	5.10	5.56	5.10	5.10	5.10	-0.14	0.12	0.25	0.37
New York, NY	16.05	16.05	16.05	14.20	NA	-0.01	0.13	1.54	NA
Philadelphia, PA	13.92	13.92	13.92	12.65	12.65	0.15	0.27	0.59	1.00
Sacramento, CA	7.96	7.55	7.55	7.34	7.13	0.31	0.15	0.29	0.46

NA: for NYC, the model-based adjustment methodology was unable to estimate ozone distributions which would meet the lower alternative standard level of 60 ppb.

Table 7C-3 Sensitivity Analysis – *ST Mortality: Regional Bayes Adjustment* (2009)
(incidence, percent of baseline mortality, incidence per 100,000) - compare with Core Results in Table 7B-2)

Study Area	Air Quality Scenario								
	Absolute Ozone-Attributable Incidence					Change in Ozone-Attributable Incidence			
	Base	75ppb	70ppb	65ppb	60ppb	Base-75	75-70	75-65	75-60
Atlanta, GA	260	260	250	240	240	5	9	17	25
	(-210 - 730)	(-210 - 710)	(-200 - 690)	(-190 - 670)	(-190 - 650)	(-4 - 13)	(-7 - 25)	(-14 - 48)	(-19 - 69)
Baltimore, MD	910	900	880	860	830	13	17	39	63
	(380 - 1400)	(370 - 1400)	(360 - 1400)	(350 - 1300)	(340 - 1300)	(5 - 20)	(7 - 26)	(16 - 62)	(26 - 100)
Boston, MA	980	990	990	980	950	-7	-3	16	43
	(330 - 1600)	(330 - 1600)	(330 - 1600)	(330 - 1600)	(320 - 1600)	(-2 - -12)	(-1 - -6)	(5 - 26)	(14 - 72)
Cleveland, OH	470	480	470	450	420	-6	14	35	59
	(110 - 830)	(120 - 840)	(110 - 810)	(110 - 780)	(100 - 740)	(-1 - -10)	(3 - 25)	(8 - 61)	(14 - 100)
Denver, CO	15	15	15	14	13	0	0	0	1
	(-330 - 350)	(-340 - 350)	(-340 - 350)	(-330 - 340)	(-310 - 320)	(1 - -2)	(-1 - 1)	(-8 - 9)	(-28 - 30)
Detroit, MI	640	640	670	650	630	NA	-24	-7	17
	(180 - 1100)	(180 - 1100)	(190 - 1100)	(180 - 1100)	(180 - 1100)	NA	(-7 - -41)	(-2 - -12)	(5 - 29)
Houston, TX	540	590	590	590	580	-46	-1	3	12
	(84 - 1000)	(91 - 1100)	(91 - 1100)	(90 - 1100)	(89 - 1100)	(-7 - -85)	(0 - -1)	(0 - 6)	(2 - 22)
Los Angeles, CA	1000	1200	1100	1100	1000	-150	38	81	150
	(-510 - 2500)	(-580 - 2900)	(-560 - 2800)	(-540 - 2700)	(-510 - 2500)	(74 - -370)	(-19 - 94)	(-40 - 200)	(-74 - 370)
New York, NY	2800	2900	2800	2500	NA	-86	93	480	NA
	(2000 - 3700)	(2000 - 3800)	(2000 - 3700)	(1700 - 3200)	NA	(-58 - -110)	(63 - 120)	(330 - 640)	NA
Philadelphia, PA	1600	1600	1600	1500	1500	-8	27	63	98
	(860 - 2300)	(870 - 2300)	(850 - 2300)	(840 - 2200)	(820 - 2200)	(-4 - -12)	(15 - 40)	(34 - 92)	(53 - 140)
Sacramento, CA	130	130	130	130	120	4	2	4	8
	(-670 - 920)	(-650 - 890)	(-640 - 880)	(-630 - 860)	(-620 - 840)	(-18 - 25)	(-11 - 15)	(-22 - 30)	(-37 - 52)
St. Louis, MO	480	480	470	450	430	2	10	26	46
	(90 - 860)	(90 - 850)	(88 - 840)	(85 - 810)	(81 - 770)	(0 - 3)	(2 - 19)	(5 - 47)	(9 - 83)

Study Area	Air Quality Scenario								
	Percent of Baseline Incidence Attributable to Ozone					Change in O_3-Attributable Risk			
	Base	75ppb	70ppb	65ppb	60ppb	Base-75	75-70	75-65	75-60
Atlanta, GA	1.1	1.1	1.0	1.0	1.0	2	3	7	9
Baltimore, MD	4.0	3.9	3.9	3.8	3.7	1	2	4	7
Boston, MA	3.4	3.5	3.5	3.4	3.3	-1	0	1	4
Cleveland, OH	2.7	2.8	2.7	2.6	2.4	-1	3	7	12
Denver, CO	0.1	0.1	0.1	0.1	0.1	0	0	2	7
Detroit, MI	3.0	3.0	3.1	3.0	2.9	NA	-4	-1	3
Houston, TX	1.5	1.6	1.6	1.6	1.6	-8	0	0	2
Los Angeles, CA	0.8	1.0	0.9	0.9	0.8	-15	3	7	13
New York, NY	4.2	4.4	4.2	3.7	NA	-3	3	16	NA
Philadelphia, PA	4.2	4.2	4.2	4.1	4.0	-1	2	4	6
Sacramento, CA	0.4	0.4	0.4	0.4	0.4	2	2	3	6
St. Louis, MO	2.8	2.8	2.8	2.7	2.6	0	2	5	9

Study Area	Air Quality Scenario								
	Ozone-Attributable Deaths per 100,000					Change in Ozone-Attributable Deaths per 100,000			
	Base	75ppb	70ppb	65ppb	60ppb	Base-75	75-70	75-65	75-60
Atlanta, GA	4.34	4.34	4.17	4.01	4.01	0.08	0.15	0.28	0.42
Baltimore, MD	16.74	16.55	16.18	15.82	15.26	0.24	0.31	0.72	1.16
Boston, MA	12.97	13.11	13.11	12.97	12.58	-0.09	-0.04	0.21	0.57
Cleveland, OH	13.81	14.10	13.81	13.22	12.34	-0.17	0.41	1.03	1.73
Denver, CO	0.40	0.40	0.40	0.38	0.35	0.00	0.00	0.01	0.03
Detroit, MI	11.39	11.39	11.92	11.56	11.21	NA	-0.43	-0.12	0.30
Houston, TX	8.39	9.16	9.16	9.16	9.01	-0.71	-0.01	0.05	0.19
Los Angeles, CA	4.63	5.56	5.10	5.10	4.63	-0.69	0.18	0.38	0.69
New York, NY	17.28	17.90	17.28	15.43	NA	-0.53	0.57	2.96	NA
Philadelphia, PA	20.24	20.24	20.24	18.98	18.98	-0.10	0.34	0.80	1.24
Sacramento, CA	2.72	2.72	2.72	2.72	2.52	0.08	0.05	0.09	0.16
St. Louis, MO	14.24	14.24	13.95	13.35	12.76	0.05	0.30	0.77	1.37

NA: for NYC, the model-based adjustment methodology was unable to estimate ozone distributions which would meet the lower alternative standard level of 60 ppb.

Table 7C-4 Sensitivity Analysis – *ST Mortality: Copollutant model (PM$_{10}$)* (2009) (incidence, percent of baseline mortality, incidence per 100,000) - compare with Core Results in Table 7B-2)

| Study Area | Air Quality Scenario | | | | | | | | |
| | Absolute Ozone-Attributable Incidence | | | | | Change in Ozone-Attributable Incidence | | | |
	Base	75ppb	70ppb	65ppb	60ppb	Base-75	75-70	75-65	75-60
Atlanta, GA	120	120	110	110	110	2	4	8	11
	(-1000 - 1200)	(-980 - 1200)	(-940 - 1100)	(-910 - 1100)	(-880 - 1100)	(-18 - 22)	(-34 - 42)	(-64 - 78)	(-91 - 110)
Baltimore, MD	460	450	450	440	420	7	8	20	32
	(-570 - 1400)	(-560 - 1400)	(-550 - 1400)	(-540 - 1400)	(-520 - 1300)	(-8 - 20)	(-10 - 27)	(-24 - 62)	(-39 - 100)
Boston, MA	180	180	180	180	170	-1	-1	3	8
	(-1100 - 1400)	(-1100 - 1400)	(-1100 - 1400)	(-1100 - 1400)	(-1100 - 1400)	(8 - -10)	(4 - -5)	(-18 - 23)	(-48 - 63)
Cleveland, OH	330	330	320	310	290	-4	10	24	40
	(-280 - 900)	(-280 - 920)	(-270 - 890)	(-260 - 850)	(-250 - 810)	(3 - -11)	(-8 - 27)	(-20 - 67)	(-34 - 110)
Denver, CO	-19	-19	-19	-19	-18	0	0	0	-2
	(-550 - 480)	(-550 - 490)	(-550 - 480)	(-540 - 470)	(-510 - 450)	(2 - -2)	(-2 - 2)	(-13 - 12)	(-46 - 42)
Detroit, MI	260	260	270	260	250	NA	-9	-3	7
	(-470 - 960)	(-470 - 960)	(-490 - 1000)	(-480 - 980)	(-460 - 940)	NA	(17 - -36)	(5 - -10)	(-12 - 26)
Houston, TX	810	880	880	870	860	-69	-1	5	18
	(-100 - 1700)	(-110 - 1800)	(-110 - 1800)	(-110 - 1800)	(-110 - 1800)	(9 - -150)	(0 - -2)	(-1 - 10)	(-2 - 38)
Los Angeles, CA	270	310	300	290	270	-39	10	21	39
	(-3300 - 3700)	(-3800 - 4300)	(-3700 - 4100)	(-3500 - 4000)	(-3300 - 3700)	(470 - -560)	(-120 - 140)	(-260 - 300)	(-470 - 550)
New York, NY	1100	1200	1100	980	NA	-33	36	190	NA
	(-850 - 3000)	(-880 - 3100)	(-850 - 3000)	(-730 - 2600)	NA	(25 - -92)	(-27 - 99)	(-140 - 520)	NA
Philadelphia, PA	850	850	840	820	800	-4	15	34	52
	(-760 - 2400)	(-760 - 2400)	(-750 - 2300)	(-730 - 2300)	(-710 - 2200)	(4 - -13)	(-13 - 42)	(-29 - 96)	(-46 - 150)
Sacramento, CA	350	340	340	330	320	10	6	12	20
	(-1000 - 1600)	(-980 - 1600)	(-960 - 1600)	(-950 - 1600)	(-920 - 1500)	(-28 - 45)	(-16 - 28)	(-32 - 55)	(-56 - 94)
St. Louis, MO	260	260	250	240	230	1	6	14	25
	(-570 - 1000)	(-570 - 1000)	(-560 - 1000)	(-540 - 990)	(-510 - 940)	(-2 - 4)	(-12 - 23)	(-30 - 57)	(-54 - 100)

| Study Area | Air Quality Scenario | | | | | | | | |
| | Percent of Baseline Incidence Attributable to Ozone | | | | | Change in O$_3$-Attributable Risk | | | |
	Base	75ppb	70ppb	65ppb	60ppb	Base-75	75-70	75-65	75-60
Atlanta, GA	0.46	0.45	0.44	0.42	0.41	0.01	0.02	0.03	0.04
Baltimore, MD	2.00	1.98	1.94	1.90	1.84	0.03	0.04	0.09	0.14
Boston, MA	0.60	0.60	0.60	0.59	0.58	0.00	0.00	0.01	0.03
Cleveland, OH	1.86	1.89	1.83	1.76	1.66	-0.02	0.06	0.14	0.23
Denver, CO	-0.23	-0.23	-0.23	-0.22	-0.21	0.00	0.00	0.00	-0.02
Detroit, MI	1.19	1.19	1.24	1.21	1.16	NA	-0.04	-0.01	0.03
Houston, TX	2.22	2.41	2.41	2.40	2.36	-0.19	0.00	0.01	0.05
Los Angeles, CA	0.21	0.24	0.23	0.22	0.21	-0.03	0.01	0.02	0.03
New York, NY	1.68	1.73	1.67	1.45	NA	-0.05	0.05	0.28	NA
Philadelphia, PA	2.23	2.25	2.21	2.16	2.11	-0.01	0.04	0.09	0.14
Sacramento, CA	1.12	1.10	1.08	1.06	1.04	0.03	0.02	0.04	0.06
St. Louis, MO	1.48	1.48	1.45	1.40	1.34	0.01	0.03	0.08	0.14

| Study Area | Air Quality Scenario | | | | | | | | |
| | Ozone-Attributable Deaths per 100,000 | | | | | Change in Ozone-Attributable Deaths per 100,000 | | | |
	Base	75ppb	70ppb	65ppb	60ppb	Base-75	75-70	75-65	75-60
Atlanta, GA	2.00	2.00	1.84	1.84	1.84	0.04	0.07	0.13	0.18
Baltimore, MD	8.46	8.28	8.28	8.09	7.72	0.12	0.15	0.37	0.59
Boston, MA	2.38	2.38	2.38	2.38	2.25	-0.02	-0.01	0.04	0.10
Cleveland, OH	9.69	9.69	9.40	9.11	8.52	-0.12	0.28	0.70	1.17
Denver, CO	-0.51	-0.51	-0.51	-0.51	-0.48	0.00	0.00	-0.01	-0.04
Detroit, MI	4.63	4.63	4.80	4.63	4.45	NA	-0.17	-0.05	0.12
Houston, TX	12.58	13.67	13.67	13.51	13.36	-1.07	-0.01	0.07	0.28
Los Angeles, CA	1.25	1.44	1.39	1.34	1.25	-0.18	0.05	0.10	0.18
New York, NY	6.79	7.41	6.79	6.05	NA	-0.20	0.22	1.17	NA
Philadelphia, PA	10.75	10.75	10.63	10.37	10.12	-0.05	0.19	0.43	0.66
Sacramento, CA	7.34	7.13	7.13	6.92	6.71	0.20	0.12	0.25	0.42
St. Louis, MO	7.72	7.72	7.42	7.12	6.83	0.03	0.16	0.42	0.74

NA: for NYC, the model-based adjustment methodology was unable to estimate ozone distributions which would meet the lower alternative standard level of 60 ppb.

Table 7C-5 Sensitivity Analysis – *ST Mortality: Alternate risk model (Zanobetti and Schwartz, 2008)* **(2009)** (incidence, percent of baseline mortality, incidence per 100,000) - compare with Core Results in Table 7B-2)

Study Area	Air Quality Scenario								
	Absolute Ozone-Attributable Incidence					Change in Ozone-Attributable Incidence			
	Base	75ppb	70ppb	65ppb	60ppb	Base-75	75-70	75-65	75-60
Atlanta, GA	140	130	120	110	110	8	10	17	23
	(-130 - 400)	(-120 - 380)	(-110 - 350)	(-110 - 330)	(-100 - 310)	(-7 - 22)	(-9 - 29)	(-16 - 49)	(-21 - 67)
Baltimore, MD	260	230	230	220	210	26	9	19	29
	(-52 - 560)	(-47 - 510)	(-45 - 490)	(-43 - 470)	(-41 - 450)	(-5 - 56)	(-2 - 19)	(-4 - 41)	(-6 - 63)
Boston, MA	380	380	370	360	350	1	10	20	33
	(21 - 730)	(21 - 730)	(20 - 710)	(20 - 690)	(19 - 670)	(0 - 3)	(1 - 19)	(1 - 39)	(2 - 64)
Cleveland, OH	190	180	180	170	150	9	8	20	33
	(-33 - 410)	(-32 - 390)	(-30 - 380)	(-28 - 350)	(-26 - 330)	(-1 - 18)	(-1 - 18)	(-3 - 43)	(-6 - 70)
Denver, CO	91	91	88	85	76	1	3	6	15
	(-120 - 290)	(-120 - 290)	(-120 - 280)	(-110 - 270)	(-99 - 240)	(-1 - 2)	(-4 - 9)	(-8 - 21)	(-20 - 49)
Detroit, MI	470	470	480	470	450	NA	-8	7	27
	(170 - 770)	(170 - 770)	(170 - 780)	(170 - 760)	(160 - 730)	NA	(-3 - -12)	(2 - 11)	(10 - 45)
Houston, TX	61	65	65	64	62	-4	0	1	3
	(-130 - 250)	(-140 - 270)	(-140 - 270)	(-140 - 260)	(-140 - 260)	(9 - -17)	(0 - 1)	(-2 - 4)	(-6 - 12)
Los Angeles, CA	460	450	430	400	380	11	21	46	67
	(-250 - 1100)	(-240 - 1100)	(-230 - 1100)	(-220 - 1000)	(-200 - 960)	(-6 - 26)	(-11 - 54)	(-24 - 110)	(-36 - 170)
New York, NY	1300	1300	1200	910	NA	84	96	370	NA
	(790 - 1900)	(740 - 1800)	(680 - 1700)	(530 - 1300)	NA	(48 - 120)	(55 - 140)	(210 - 520)	NA
Philadelphia, PA	490	450	440	420	400	32	16	35	52
	(5 - 950)	(5 - 890)	(5 - 860)	(5 - 830)	(4 - 790)	(0 - 64)	(0 - 32)	(0 - 69)	(1 - 100)
Sacramento, CA	240	210	200	190	180	35	8	16	26
	(-85 - 560)	(-73 - 480)	(-70 - 470)	(-68 - 450)	(-64 - 430)	(-12 - 81)	(-3 - 20)	(-6 - 38)	(-9 - 62)
St. Louis, MO	190	190	180	170	160	3	10	19	30
	(-39 - 410)	(-39 - 410)	(-37 - 390)	(-35 - 370)	(-33 - 350)	(-1 - 6)	(-2 - 21)	(-4 - 42)	(-6 - 65)

Study Area	Air Quality Scenario								
	Percent of Baseline Incidence Attributable to Ozone					Change in O$_3$-Attributable Risk			
	Base	75ppb	70ppb	65ppb	60ppb	Base-75	75-70	75-65	75-60
Atlanta, GA	1.4	1.3	1.2	1.1	1.1	5	7	13	17
Baltimore, MD	2.5	2.2	2.2	2.1	2.0	10	4	8	12
Boston, MA	2.5	2.5	2.4	2.4	2.3	0	2	5	8
Cleveland, OH	2.5	2.4	2.3	2.1	1.9	4	4	10	17
Denver, CO	1.8	1.8	1.8	1.7	1.5	1	3	7	16
Detroit, MI	4.1	4.1	4.2	4.1	3.9	NA	-2	1	5
Houston, TX	0.6	0.6	0.6	0.6	0.6	-7	0	2	4
Los Angeles, CA	1.4	1.4	1.3	1.2	1.2	2	5	10	15
New York, NY	4.5	4.2	3.9	3.0	NA	6	7	28	NA
Philadelphia, PA	2.8	2.6	2.5	2.4	2.3	6	3	7	11
Sacramento, CA	2.9	2.5	2.4	2.3	2.2	14	4	7	12
St. Louis, MO	2.4	2.4	2.3	2.2	2.0	1	5	10	15

Study Area	Air Quality Scenario								
	Ozone-Attributable Deaths per 100,000					Change in Ozone-Attributable Deaths per 100,000			
	Base	75ppb	70ppb	65ppb	60ppb	Base-75	75-70	75-65	75-60
Atlanta, GA	2.34	2.17	2.00	1.84	1.84	0.13	0.17	0.28	0.38
Baltimore, MD	4.78	4.23	4.23	4.05	3.86	0.48	0.16	0.35	0.53
Boston, MA	5.03	5.03	4.90	4.77	4.63	0.02	0.13	0.26	0.44
Cleveland, OH	5.58	5.29	5.29	4.99	4.41	0.25	0.25	0.59	0.97
Denver, CO	2.45	2.45	2.37	2.29	2.05	0.01	0.08	0.17	0.40
Detroit, MI	8.36	8.36	8.54	8.36	8.01	NA	-0.13	0.12	0.48
Houston, TX	0.95	1.01	1.01	0.99	0.96	-0.07	0.00	0.02	0.05
Los Angeles, CA	2.13	2.08	1.99	1.85	1.76	0.05	0.10	0.21	0.31
New York, NY	8.02	8.02	7.41	5.62	NA	0.52	0.59	2.28	NA
Philadelphia, PA	6.20	5.69	5.57	5.31	5.06	0.40	0.20	0.44	0.66
Sacramento, CA	5.03	4.40	4.19	3.98	3.77	0.73	0.18	0.34	0.54
St. Louis, MO	5.64	5.64	5.34	5.04	4.75	0.08	0.28	0.56	0.89

NA: for NYC, the model-based adjustment methodology was unable to estimate ozone distributions which would meet the lower alternative standard level of 60 ppb.

Table 7C-6 Sensitivity Analysis – *LT Mortality: Alternate risk model (regional effect estiamtes)* **(2009)** (incidence, percent of baseline mortality, incidence per 100,000) - compare with Core Results in Table 7B-7)

Study Area	Air Quality Scenario								
	Absolute Ozone-Attributable Incidence					Change in Ozone-Attributable Incidence			
	Base	75ppb	70ppb	65ppb	60ppb	Base-75	75-70	75-65	75-60
Atlanta, GA	1,684	1,632	1,565	1,507	1,455	90	111	204	281
	(870 - 2277)	(838 - 2217)	(799 - 2140)	(764 - 2071)	(734 - 2010)	(41 - 139)	(50 - 171)	(92 - 312)	(128 - 428)
Baltimore, MD	-214	-201	-194	-187	-179	-12	-6	-13	-21
	(-2241 - 1145)	(-2078 - 1088)	(-2000 - 1059)	(-1906 - 1024)	(-1814 - 989)	(-107 - 80)	(-52 - 39)	(-116 - 86)	(-181 - 132)
Boston, MA	-290	-291	-289	-280	-269	1	-2	-10	-21
	(-2952 - 1601)	(-2959 - 1604)	(-2938 - 1596)	(-2832 - 1554)	(-2700 - 1501)	(4 - -3)	(-14 - 11)	(-87 - 66)	(-182 - 135)
Cleveland, OH	0	0	0	0	0	0	0	0	0
	(-1592 - 1021)	(-1538 - 998)	(-1456 - 963)	(-1358 - 919)	(-1259 - 872)	(-35 - 35)	(-55 - 53)	(-122 - 117)	(-194 - 181)
Denver, CO	665	664	655	635	590	1	13	40	99
	(-40 - 1158)	(-40 - 1157)	(-40 - 1144)	(-38 - 1115)	(-35 - 1048)	(0 - 2)	(-1 - 26)	(-2 - 81)	(-5 - 198)
Detroit, MI	0	0	0	0	0	0	0	0	0
	(-2176 - 1406)	(0 - 0)	(-2207 - 1418)	(-2115 - 1380)	(-2004 - 1331)	(0 - 0)	(0 - 0)	(0 - 0)	(0 - 0)
Houston, TX	1,402	1,410	1,388	1,362	1,329	-13	37	81	134
	(720 - 1907)	(724 - 1916)	(711 - 1891)	(696 - 1861)	(676 - 1822)	(-6 - -21)	(17 - 58)	(36 - 124)	(61 - 206)
Los Angeles, CA	751	1,452	695	662	626	27	65	136	211
	(-3983 - 4116)	(-7652 - 7999)	(-3637 - 3856)	(-3434 - 3696)	(-3220 - 3522)	(-120 - 171)	(-292 - 416)	(-615 - 862)	(-964 - 1329)
New York, NY	-528	-513	-485	-413	NA	-14	-27	-96	NA
	(-5446 - 2871)	(-5268 - 2805)	(-4928 - 2676)	(-4090 - 2334)	NA	(-119 - 90)	(-232 - 173)	(-849 - 605)	NA
Philadelphia, PA	-346	-331	-320	-307	-296	-14	-11	-23	-34
	(-3604 - 1866)	(-3423 - 1801)	(-3284 - 1750)	(-3128 - 1690)	(-2991 - 1637)	(-119 - 89)	(-93 - 70)	(-201 - 149)	(-298 - 218)
Sacramento, CA	1,139	988	951	917	873	204	48	92	148
	(-70 - 1971)	(-59 - 1750)	(-56 - 1695)	(-54 - 1643)	(-51 - 1575)	(-11 - 406)	(-2 - 98)	(-5 - 187)	(-8 - 297)
St. Louis, MO	0	0	0	0	0	0	0	0	0
	(-1796 - 1130)	(-1770 - 1120)	(-1687 - 1085)	(-1590 - 1044)	(-1478 - 995)	(-16 - 16)	(-54 - 53)	(-119 - 114)	(-197 - 185)

Study Area	Air Quality Scenario								
	Percent of Baseline Incidence Attributable to Ozone					Change in O$_3$-Attributable Risk			
	Base	75ppb	70ppb	65ppb	60ppb	Base-75	75-70	75-65	75-60
Atlanta, GA	42.5	41.2	39.6	38.1	36.8	3.1	4.0	7.6	10.6
Baltimore, MD	-7.6	-7.0	-6.7	-6.4	-6.1	7.7	4.0	8.8	13.4
Boston, MA	-6.2	-6.2	-6.1	-5.9	-5.6	-0.2	0.7	4.5	9.3
Cleveland, OH	0.0	0.0	0.0	0.0	0.0	0.0	0.0	0.0	0.0
Denver, CO	27.4	27.4	27.0	26.2	24.4	0.1	1.3	4.2	10.7
Detroit, MI	0.0	0.0	0.0	0.0	0.0	0.0	0.0	0.0	0.0
Houston, TX	40.9	41.1	40.5	39.8	38.8	-0.5	1.5	3.4	5.7
Los Angeles, CA	4.6	8.9	4.3	4.2	4.0	-95.3	51.5	53.2	55.2
New York, NY	-6.8	-6.6	-6.2	-5.0	NA	3.5	6.8	23.6	NA
Philadelphia, PA	-7.3	-6.9	-6.6	-6.3	-6.0	5.3	4.3	9.2	13.4
Sacramento, CA	28.4	24.8	23.9	23.1	22.0	12.8	3.6	6.9	11.2
St. Louis, MO	0.0	0.0	0.0	0.0	0.0	0.0	0.0	0.0	0.0

Study Area	Air Quality Scenario								
	Ozone-Attributable Deaths per 100,000					Change in Ozone-Attributable Deaths per 100,000			
	Base	75ppb	70ppb	65ppb	60ppb	Base-75	75-70	75-65	75-60
Atlanta, GA	49.25	47.72	45.77	44.06	42.57	2.64	3.26	5.97	8.23
Baltimore, MD	-6.56	-6.16	-5.97	-5.73	-5.50	-0.38	-0.19	-0.41	-0.64
Boston, MA	-6.27	-6.28	-6.25	-6.05	-5.81	0.01	-0.04	-0.22	-0.45
Cleveland, OH	0.00	0.00	0.00	0.00	0.00	0.00	0.00	0.00	0.00
Denver, CO	31.10	31.06	30.63	29.71	27.61	0.05	0.59	1.87	4.63
Detroit, MI	0.00	0.00	0.00	0.00	0.00	0.00	0.00	0.00	0.00
Houston, TX	39.74	39.96	39.34	38.59	37.66	-0.38	1.05	2.29	3.80
Los Angeles, CA	6.24	6.03	5.78	5.50	5.20	0.22	0.27	0.56	0.88
New York, NY	-5.38	-5.23	-4.94	-4.21	NA	-0.14	-0.27	-0.98	NA
Philadelphia, PA	-7.32	-7.01	-6.77	-6.50	-6.26	-0.29	-0.23	-0.49	-0.72
Sacramento, CA	41.19	35.71	34.39	33.17	31.57	7.38	1.75	3.34	5.36
St. Louis, MO	0.00	0.00	0.00	0.00	0.00	0.00	0.00	0.00	0.00

NA: for NYC, the model-based adjustment methodology was unable to estimate ozone distributions which would meet the lower alternative standard level of 60 ppb.

Table 7C-7 Sensitivity Analysis – *LT Mortality: Alternate risk model (ozone-only effect estimate)* (2009) (incidence, percent of baseline mortality, incidence per 100,000) - compare with Core Results in Table 7B-7)

Study Area	Air Quality Scenario								
	Absolute Ozone-Attributable Incidence					Change in Ozone-Attributable Incidence			
	Base	75ppb	70ppb	65ppb	60ppb	Base-75	75-70	75-65	75-60
Atlanta, GA	200	200	190	180	170	9.0	11	21	29
	(61 - 330)	(58 - 320)	(55 - 310)	(53 - 290)	(50 - 280)	(3 - 15)	(3 - 19)	(6 - 35)	(8 - 49)
Baltimore, MD	220	200	200	190	180	14	6.7	15	23
	(65 - 350)	(61 - 330)	(59 - 320)	(57 - 310)	(55 - 300)	(4 - 23)	(2 - 11)	(4 - 25)	(7 - 39)
Boston, MA	300	300	300	290	280	-0 57	1.8	11	23
	(89 - 490)	(89 - 490)	(88 - 490)	(86 - 470)	(82 - 460)	(0 - -1)	(1 - 3)	(3 - 19)	(7 - 40)
Cleveland, OH	150	140	140	130	120	4.3	6.7	15	23
	(44 - 240)	(43 - 230)	(41 - 220)	(39 - 210)	(36 - 200)	(1 - 7)	(2 - 11)	(4 - 25)	(7 - 39)
Denver, CO	140	140	140	130	120	0.21	2.4	7.7	19
	(42 - 230)	(42 - 230)	(41 - 220)	(40 - 220)	(37 - 200)	(0 - 0)	(1 - 4)	(2 - 13)	(6 - 33)
Detroit, MI	200	200	200	200	190	NA	-2.4	4.9	14
	(60 - 330)	(60 - 330)	(61 - 330)	(59 - 320)	(56 - 310)	NA	(-1 - -4)	(1 - 8)	(4 - 24)
Houston, TX	170	170	170	160	160	-1.3	3.7	8.0	13
	(50 - 280)	(50 - 280)	(49 - 270)	(48 - 270)	(47 - 260)	(0 - -2)	(1 - 6)	(2 - 14)	(4 - 23)
Los Angeles, CA	800	780	740	710	670	30	36	76	120
	(240 - 1300)	(230 - 1300)	(220 - 1200)	(210 - 1200)	(200 - 1100)	(9 - 51)	(10 - 62)	(22 - 130)	(34 - 200)
New York, NY	540	520	500	430	NA	15	30	110	NA
	(160 - 880)	(160 - 860)	(150 - 820)	(130 - 710)	NA	(4 - 26)	(9 - 51)	(30 - 180)	NA
Philadelphia, PA	350	340	330	310	300	15	12	26	38
	(100 - 570)	(100 - 550)	(97 - 540)	(94 - 520)	(90 - 500)	(4 - 26)	(3 - 21)	(7 - 44)	(11 - 64)
Sacramento, CA	240	200	200	190	180	40	9.3	18	29
	(72 - 390)	(61 - 330)	(59 - 320)	(56 - 310)	(53 - 290)	(11 - 67)	(3 - 16)	(5 - 30)	(8 - 49)
St. Louis, MO	160	160	160	150	140	2.0	6.6	14	23
	(49 - 270)	(48 - 260)	(46 - 250)	(44 - 240)	(42 - 230)	(1 - 3)	(49 - 270)	(4 - 24)	(7 - 40)

Study Area	Air Quality Scenario								
	Percent of Baseline Incidence Attributable to Ozone					Change in O_3-Attributable Risk			
	Base	75ppb	70ppb	65ppb	60ppb	Base-75	75-70	75-65	75-60
Atlanta, GA	12.4	11.9	11.3	10.8	10.3	4	5	10	14
Baltimore, MD	12.9	12.2	11.8	11.4	11.0	6	3	7	10
Boston, MA	11.1	11.1	11.1	10.8	10.4	0	1	3	7
Cleveland, OH	12.1	11.8	11.3	10.7	10.1	3	4	10	15
Denver, CO	14.1	14.1	13.9	13.4	12.4	0	2	5	12
Detroit, MI	11.9	11.9	12.0	11.6	11.2	NA	-1	2	6
Houston, TX	11.8	11.9	11.6	11.4	11.0	-1	2	4	7
Los Angeles, CA	15.1	14.6	14.0	13.4	12.7	3	4	9	14
New York, NY	12.0	11.7	11.1	9.6	NA	3	5	19	NA
Philadelphia, PA	12.5	12.0	11.7	11.2	10.9	4	3	7	10
Sacramento, CA	14.7	12.6	12.1	11.6	11.1	14	4	8	13
St. Louis, MO	12.6	12.4	12.0	11.5	10.8	1	4	8	13

Study Area	Air Quality Scenario								
	Ozone-Attributable Deaths per 100,000					Change in Ozone-Attributable Deaths per 100,000			
	Base	75ppb	70ppb	65ppb	60ppb	Base-75	75-70	75-65	75-60
Atlanta, GA	5.95	5.72	5.43	5.19	4.97	0.26	0.32	0.60	0.83
Baltimore, MD	6.63	6.26	6.08	5.86	5.64	0.42	0.21	0.45	0.70
Boston, MA	6.42	6.43	6.40	6.22	5.99	-0 01	0.04	0.24	0.50
Cleveland, OH	6.96	6.78	6.50	6.16	5.80	0.21	0.32	0.70	1.09
Denver, CO	6.51	6.50	6.40	6.19	5.71	0.01	0.11	0.36	0.90
Detroit, MI	5.97	5.97	6.04	5.84	5.61	0.00	-0 07	0.15	0.41
Houston, TX	4.76	4.79	4.70	4.59	4.45	-0 04	0.10	0.23	0.38
Los Angeles, CA	6.66	6.44	6.19	5.90	5.59	0.25	0.30	0.63	0.98
New York, NY	5.47	5.33	5.06	4.36	NA	0.16	0.30	1.07	NA
Philadelphia, PA	7.42	7.13	6.91	6.65	6.42	0.32	0.25	0.54	0.80
Sacramento, CA	8.65	7.40	7.10	6.83	6.48	1.44	0.34	0.65	1.04
St. Louis, MO	8.05	7.96	7.68	7.34	6.94	0.10	0.33	0.71	1.15

NA: for NYC, the model-based adjustment methodology was unable to estimate ozone distributions which would meet the lower alternative standard level of 60 ppb.

APPENDIX 8-A. CITY-SPECIFIC OZONE-MORTALITY EFFECT ESTIMATES

Table 8-A-1. Smith et al. (2009) city-specific and regional non-accidental mortality effect estimates for 8-hr daily maximum ozone, using April-October (many just May-September) ozone observations from 1987-2000, based on 98 U.S. urban communities.

Location	National prior		Regional prior	
	Beta	Std	Beta	Std
Akron, OH	0.000305	0.000332	0.000502	0.000279
Albuquerque, NM	0.000292	0.000349	-5E-05	0.000351
Arlington, VA	0.000341	0.000353	0.00091	0.000297
Atlanta, GA	0.000256	0.000283	0.000222	0.000229
Austin, TX	0.000309	0.000313	-1.6E-05	0.000326
Bakersfield, CA	0.000342	0.00033	4.41E-05	0.000282
Baltimore, MD	0.000313	0.000322	0.000863	0.000286
Baton Rouge, LA	0.000383	0.00031	0.000281	0.000244
Biddeford, ME	0.000321	0.000349	0.000897	0.000297
Birmingham, AL	0.000212	0.000318	0.000197	0.00025
Boston, MA	0.000156	0.000343	0.000803	0.00031
Buffalo, NY	0.000349	0.000324	0.00052	0.000274
Cedar Rapids, IA	0.000338	0.000338	-0.00017	0.000482
Charlotte, NC	0.000236	0.000328	0.000208	0.000254
Chicago, IL	0.000498	0.000247	0.000568	0.000224
Cincinnati, OH	0.000513	0.000329	0.000597	0.000277
Cleveland, OH	0.000488	0.000308	0.00058	0.000264
Colorado Springs, CO	0.000389	0.000347	0.000257	0.000377
Columbus, GA	0.000405	0.000321	0.000289	0.000249
Columbus, OH	0.000309	0.000323	0.000501	0.000274
Corpus Christi, TX	0.000375	0.000322	8.6E-06	0.000335
Coventry, RI	0.000251	0.000335	0.000847	0.000297
Dallas/Ft Worth, TX	0.000538	0.000238	0.000392	0.000213
Dayton, OH	0.000314	0.000334	0.000507	0.00028
Denver, CO	0.000182	0.000345	0.000163	0.000372
Des Moines, IA	0.000188	0.000342	-0.00024	0.00048
Detroit, MI	0.000439	0.000299	0.000554	0.000259
El Paso, TX	0.000173	0.000347	-9.7E-05	0.000347
Evansville	0.000275	0.000326	0.000486	0.000277
Ft Wayne, IN	0.000319	0.00034	0.000512	0.000283

Fresno, CA	0.000188	0.000334	-2.6E-05	0.000283
Grand Rapids, MI	0.000377	0.000335	0.000537	0.00028
Greensboro, NC	0.000291	0.000346	0.000231	0.000262
Honolulu, HI	0.000451	0.000358	6.17E-05	0.000236
Houston, TX	0.000403	0.000215	0.00032	0.000189
Huntsville, AL	0.000548	0.000357	0.000349	0.000271
Indianapolis, IN	0.000246	0.000331	0.000474	0.00028
Industrial Midwest	N/A	N/A	0.000521	0.00018
Jackson, MS	0.000269	0.000327	0.000223	0.000253
Jacksonville, FL	0.000201	0.000322	0.000192	0.000252
Jersey City, NJ	0.000117	0.000336	0.000769	0.000314
Johnston, PA	0.000329	0.000334	0.000884	0.00029
Kansas City, KS	0.000146	0.000339	-0.00025	0.000473
Kansas City, MO	0.000395	0.000321	-0.00013	0.000467
Kingston, NY	0.000452	0.000342	0.000944	0.000287
Knoxville, TN	0.000471	0.000339	0.000316	0.00026
Lafayette, LA	0.000236	0.000315	0.000209	0.000247
Lake Charles, LA	0.000263	0.000312	0.000222	0.000245
Las Vegas, NV	0.00014	0.000348	-0.00011	0.000346
Lexington, KY	0.000172	0.000345	0.000443	0.00029
Lincoln, NE	0.000426	0.000349	0.000941	0.000292
Little Rock, AR	0.000217	0.000339	0.000198	0.000261
Los Angeles, CA	0.000148	0.000165	5.24E-05	0.000161
Louisville, KY	0.000351	0.000322	0.000521	0.000273
Madison, WI	0.000456	0.000355	0.000577	0.000292
Memphis, TN	0.000391	0.000312	0.000284	0.000245
Miami, FL	0.000233	0.000277	0.000211	0.000226
Milwaukee, WI	0.00029	0.000315	0.000488	0.00027
Mobile, AL	0.000359	0.000323	0.000266	0.00025
Modesto, CA	0.000322	0.000341	0.000227	0.000372
Muskegon, MI	0.000305	0.000346	0.000508	0.000287
Nashville, TN	0.000347	0.000329	0.00026	0.000253
National Average	0.000322	8.42E-05	N/A	N/A
New Orleans, LA	0.000252	0.000321	0.000216	0.00025
New York, NY	0.000917	0.00023	0.001055	0.000195
Newark, NJ	0.000549	0.000333	0.000972	0.000276
North East	N/A	N/A	0.000908	0.000192
North West	N/A	N/A	0.000224	0.000308
Oakland, CA	0.000214	0.000334	0.000179	0.000366
Oklahoma City, OK	0.000358	0.000317	4.62E-06	0.000331
Omaha, NE	0.000377	0.000345	0.000917	0.000292

Orlando, FL	-3.3E-05	0.000358	7.91E-05	0.000286
Philadelphia, PA	0.000574	0.000296	0.000948	0.000253
Phoenix, AZ	0.00034	0.000301	7.84E-06	0.00032
Pittsburgh, PA	0.000155	0.000306	0.000412	0.000272
Portland, OR	0.00037	0.000335	0.00025	0.000369
Providence, RI	0.000418	0.000333	0.000922	0.000284
Raleigh, NC	0.000271	0.000337	0.000223	0.000258
Riverside, CA	0.000206	0.000295	2.3E-06	0.000257
Rochester, NY	0.000406	0.000339	0.000923	0.000287
Sacramento, CA	0.000306	0.000313	0.000225	0.000351
Salt Lake City, UT	0.000296	0.000345	0.000215	0.000375
San Antonio, TX	7.15E-05	0.000307	-0.00012	0.000313
San Bernardino, CA	0.00034	0.000283	7.61E-05	0.000254
San Diego, CA	0.000118	0.000289	-3.6E-05	0.000252
San Jose, CA	0.000351	0.000323	0.000244	0.00036
Santa Ana/Anaheim, CA	0.0002	0.000279	8.65E-06	0.000246
Seattle, WA	0.000283	0.000325	0.000212	0.00036
Shreveport, LA	0.000366	0.00032	0.00027	0.000248
South East	N/A	N/A	0.000242	0.000135
South West	N/A	N/A	-4.4E-05	0.000273
Southern California	N/A	N/A	1.73E-05	0.000189
Spokane, WA	0.000327	0.000353	0.000227	0.000381
St. Louis, MO	0.000476	0.000336	0.000581	0.000281
St Petersburg, FL	0.000147	0.000288	0.000166	0.000235
Stockton, CA	0.00036	0.000343	0.000244	0.000374
Syracuse, NY	0.00053	0.000357	0.000985	0.000292
Tacoma, WA	0.00036	0.000342	0.000245	0.000373
Tampa, FL	0.000223	0.000299	0.000204	0.000239
Toledo, OH	0.000414	0.000333	0.000553	0.000279
Tucson, AZ	0.000333	0.000334	-2.1E-05	0.000342
Tulsa, OK	0.000382	0.000325	0.000277	0.000251
Upper Midwest	N/A	N/A	-0.0002	0.000445
Washington, DC	0.000239	0.000321	0.000823	0.000294
Wichita, KS	0.000249	0.000345	-0.00022	0.000486
Worcester, MA	0.000467	0.000337	0.000946	0.000283

Table 8-A-2. Zanobetti and Schwartz (2008) city-specific all-cause mortality effect estimates for June-August 8-hr daily mean (10am-6pm) ozone from 1989-2000, using a 0-3 day lag, based on 48 U.S. cities.

Location	Beta	Std
All cities (48)	0.00053	0.000125
Albuquerque, NM	0.000528	0.000416
Atlanta, GA	0.000295	0.000289
Austin, TX	0.00045	0.000393
Baltimore, MD	0.000515	0.000314
Birmingham, AL	0.000293	0.000356
Boston, MA	0.000682	0.000328
Boulder, CO	0.000602	0.000419
Broward, FL	0.000593	0.000382
Canton, OH	0.000489	0.000401
Charlotte, NC	0.000571	0.000381
Chicago, IL	0.000479	0.000299
Cincinnati, OH	0.000509	0.000361
Cleveland, OH	0.000596	0.000355
Colorado Springs, CO	0.000497	0.000418
Columbus, OH	0.000739	0.000368
Dallas, TX	0.000578	0.000317
Denver, CO	0.000352	0.000409
Detroit, MI	0.001046	0.000344
Greensboro, NC	0.000478	0.000397
Honolulu, HI	0.000486	0.00042
Houston, TX	0.000163	0.000263
Jersey city, NJ	0.000354	0.00038
Kansas City, KS	0.000922	0.000387
Los Angeles, CA	0.000274	0.000213
Miami, FL	0.000607	0.000373
Milwaukee, WI	0.000659	0.000382
Nashville, TN	0.00046	0.000383
New Haven, CT	0.000647	0.000364
New Orleans, LA	0.000218	0.000375
New York, NY	0.001092	0.000236
Oklahoma City, OK	0.00062	0.00038
Orlando, FL	0.000487	0.000377
Philadelphia, PA	0.000625	0.000315
Phoenix, AZ	0.00071	0.000374
Pittsburgh, PA	0.00028	0.000328
Provo/Orem, UT	0.000527	0.00042

Sacramento, CA	0.000569	0.000389
Salt Lake City, UT	0.000478	0.000407
San Diego, CA	0.000448	0.000373
San Francisco, CA	0.000566	0.000416
Seattle, WA	0.000491	0.00038
Spokane, WA	0.00059	0.000415
St. Louis, MO	0.000544	0.000333
Tampa, FL	0.000123	0.000366
Terra Haute, IN	0.000659	0.00042
Tulsa, OK	0.000871	0.000391
Washington, DC	9.56E-05	0.00036
Youngstown, OH	0.000448	0.000394

APPENDIX 8-B. Supplement to the Representativeness Analysis of the 12 Urban Study Areas

Following the analysis discussed in Section 8.2, this appendix provides graphical comparisons of the empirical distributions of components of the risk function, and additional variables that have been identified as potentially influencing the risk associated with ozone exposures. In each graph, the blue line represents the cumulative distribution function (CDF) for the complete set of data available for the variable. In some cases, this many encompass all counties in the U.S., while in others it may be based on a subset of the U.S., usually for large urban areas. The black squares at the bottom of each graph represent the specific value of the variable for one of the case study locations, with the line showing where that value intersect the CDF of the nationwide data.

Elements of the Risk Equation

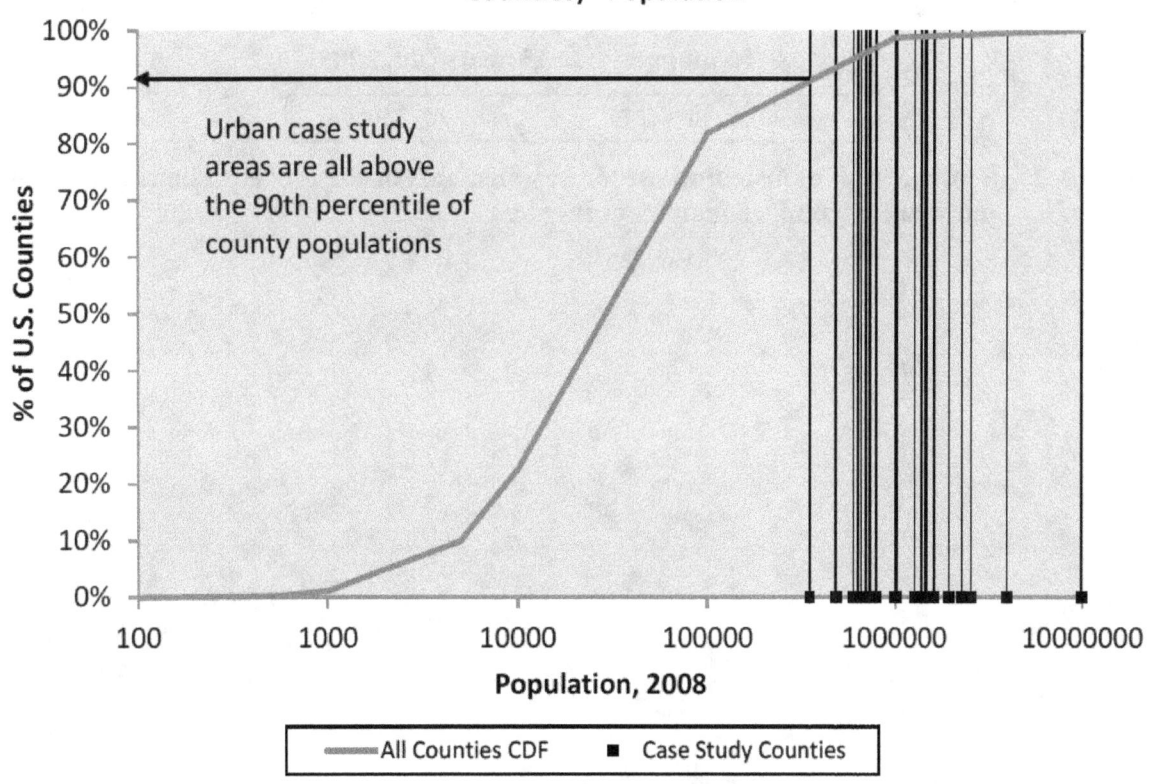

Figure A.1   Comparison of distributions for key elements of the risk equation: Total population

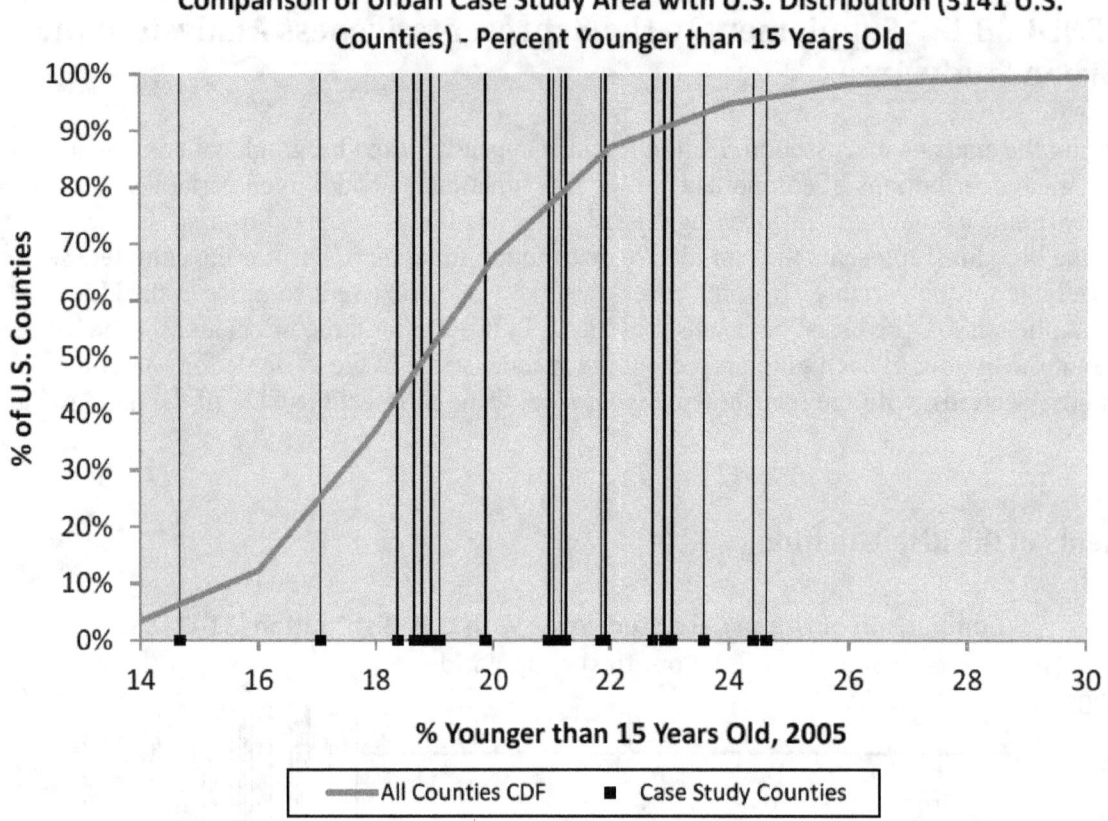

Figure A.2   Comparison of distributions for key elements of the risk equation: Percent of population younger than 15 years old

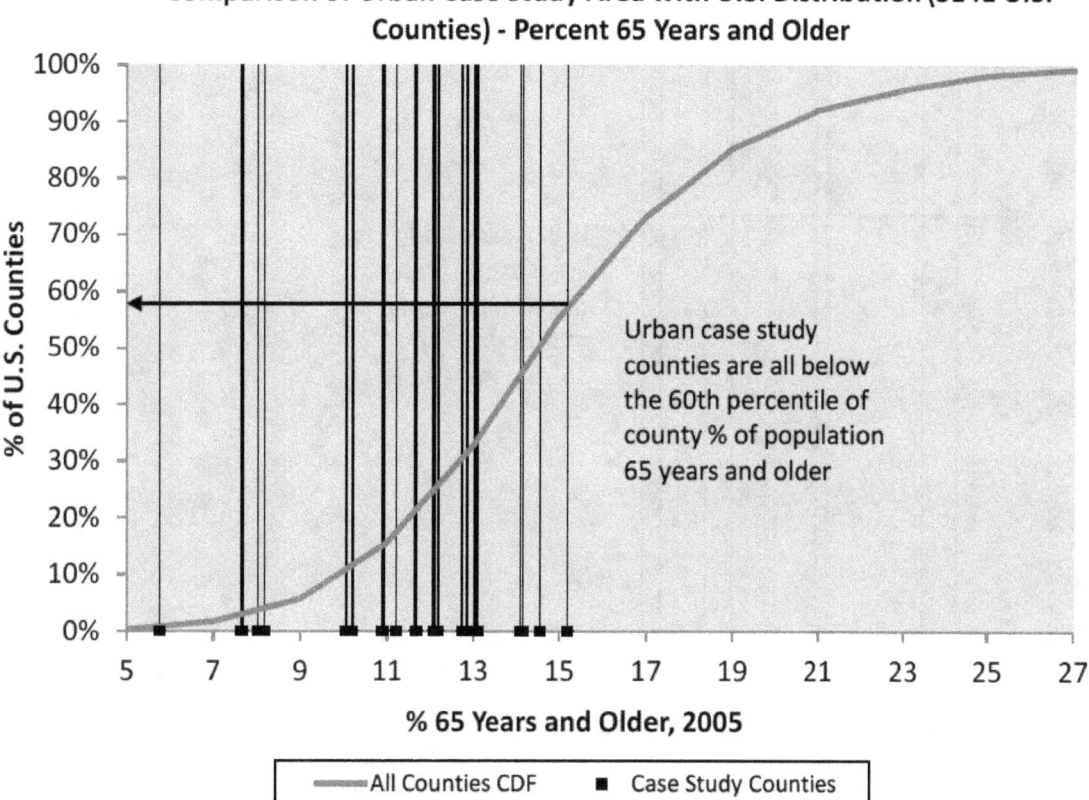

Figure A.3 Comparison of distributions for key elements of the risk equation: Percent of population 65 and older

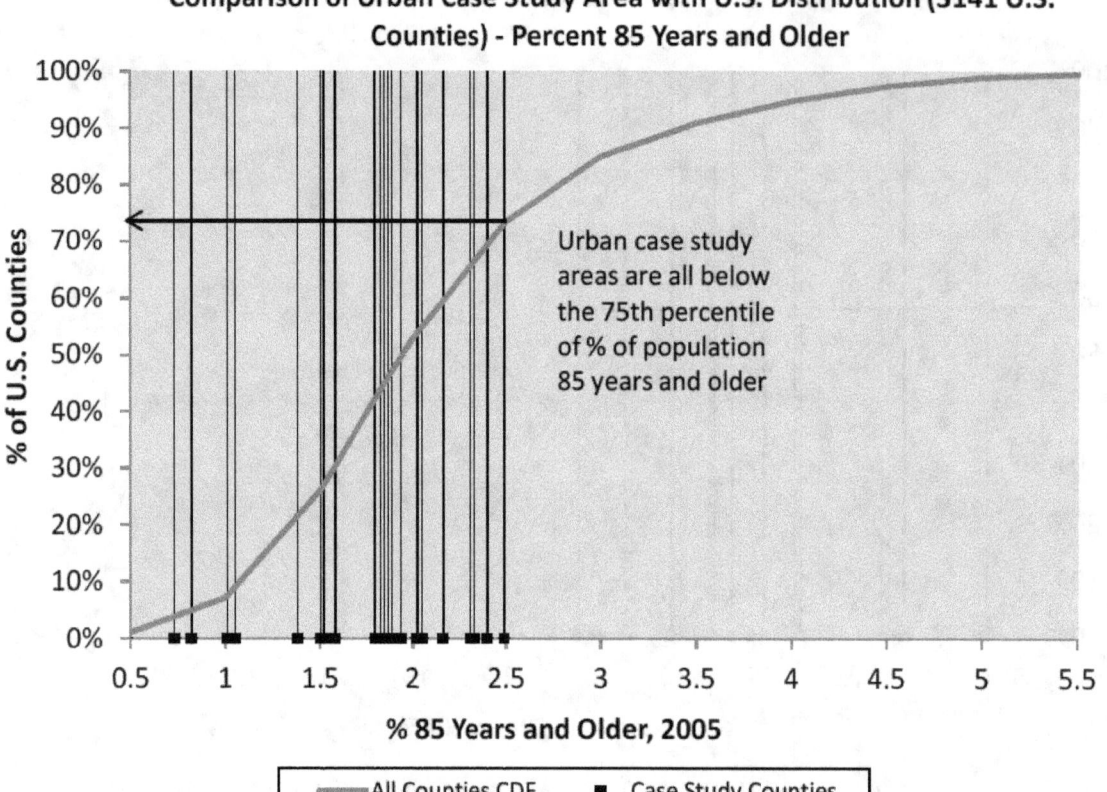

Figure A.4 Comparison of distributions for key elements of the risk equation: Percent of population 85 and older

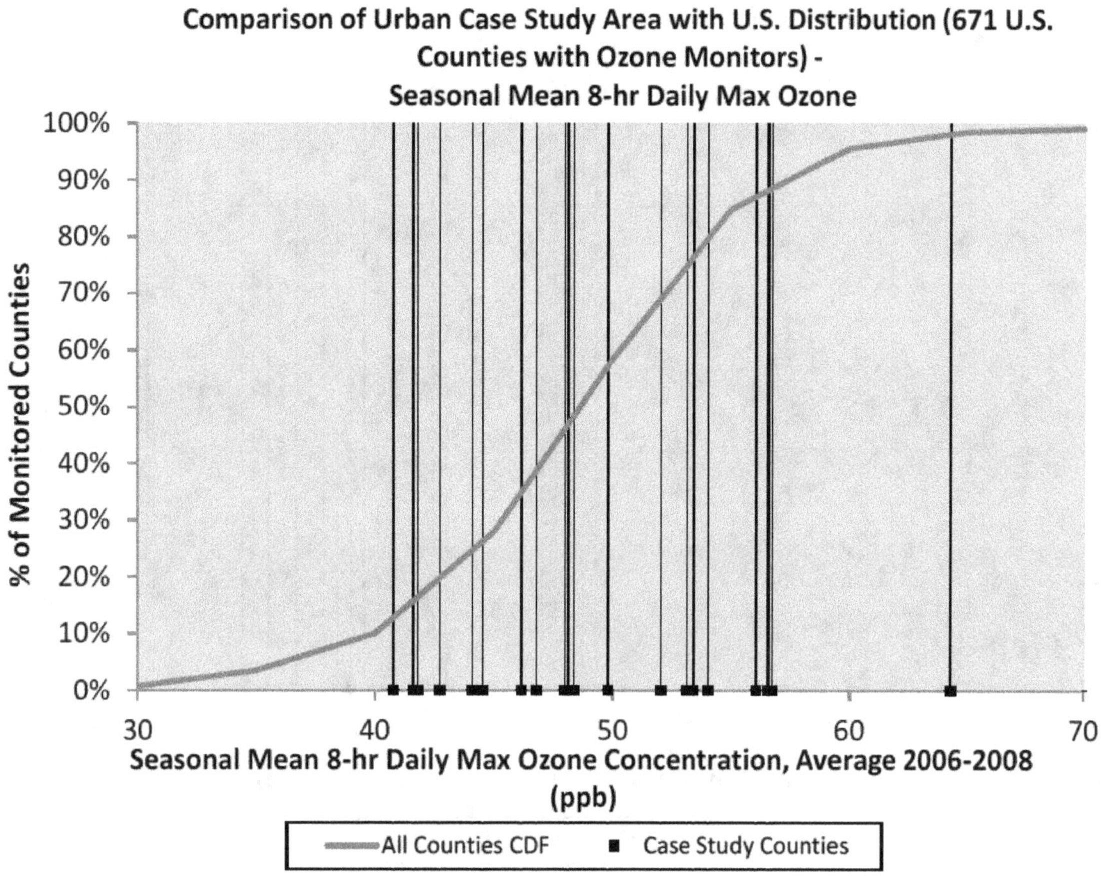

Figure A.5 Comparison of distributions for key elements of the risk equation: Seasonal mean 8-hr daily maximum ozone concentration

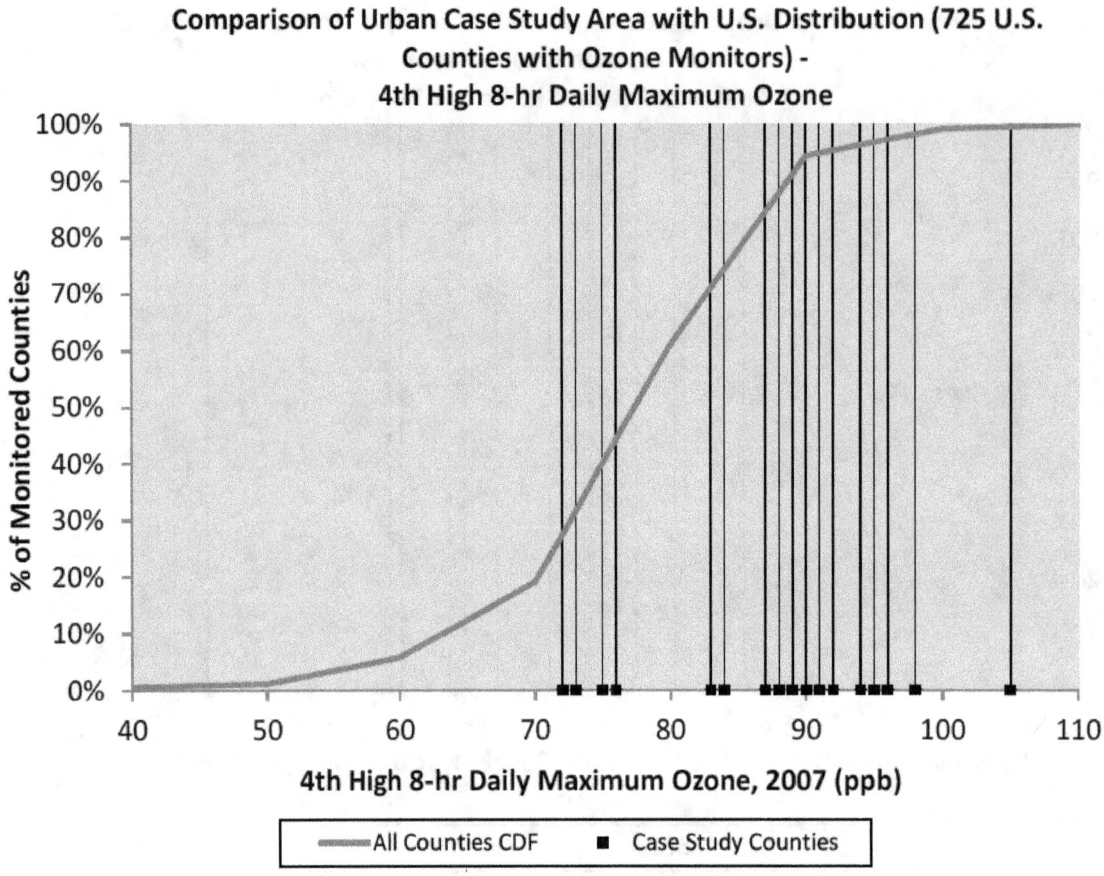

Figure A.6 Comparison of distributions for key elements of the risk equation: 4th highest 8-hr daily maximum ozone concentration

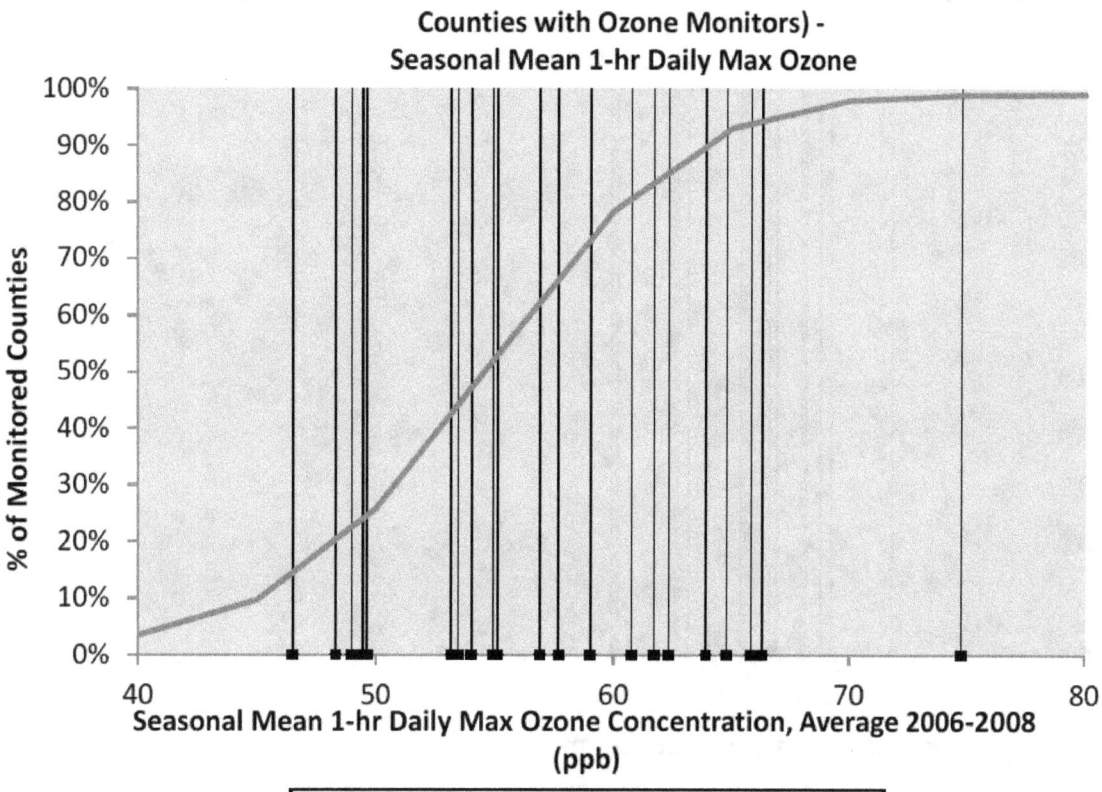

Figure A.7   Comparison of distributions for key elements of the risk equation: Seasonal mean 1-hr daily maximum ozone concentration

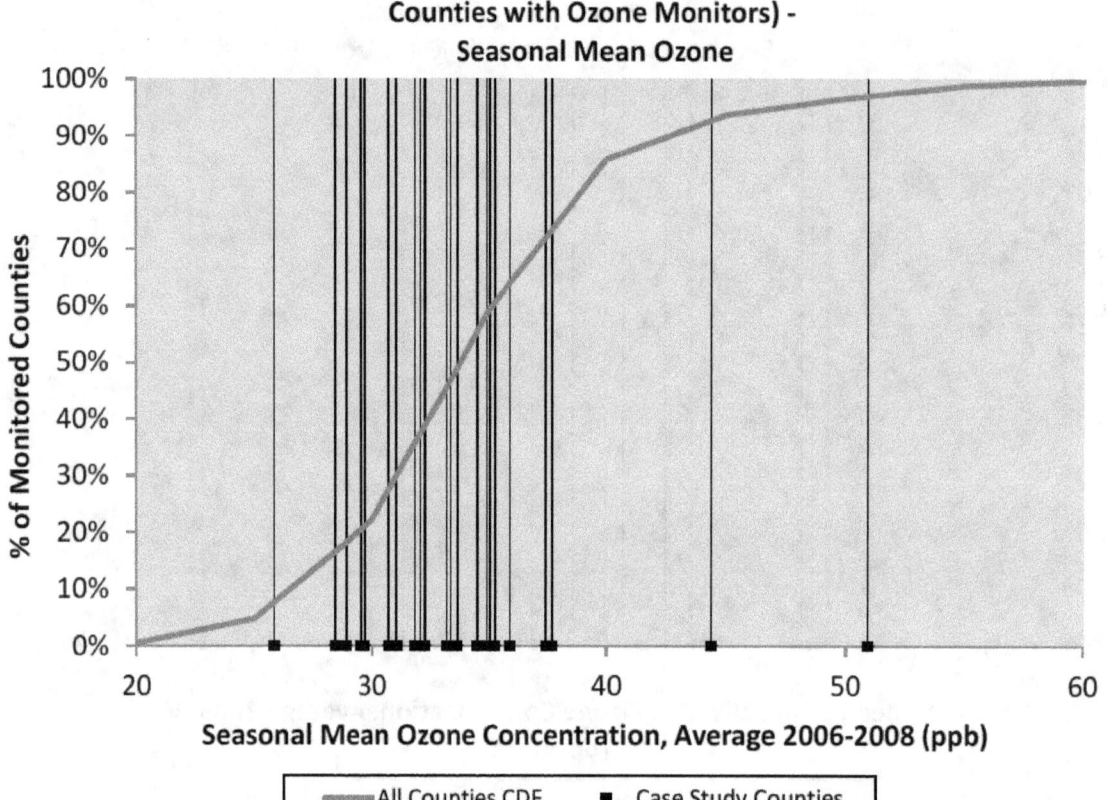

Figure A.8   **Comparison of distributions for key elements of the risk equation: Seasonal mean ozone concentration**

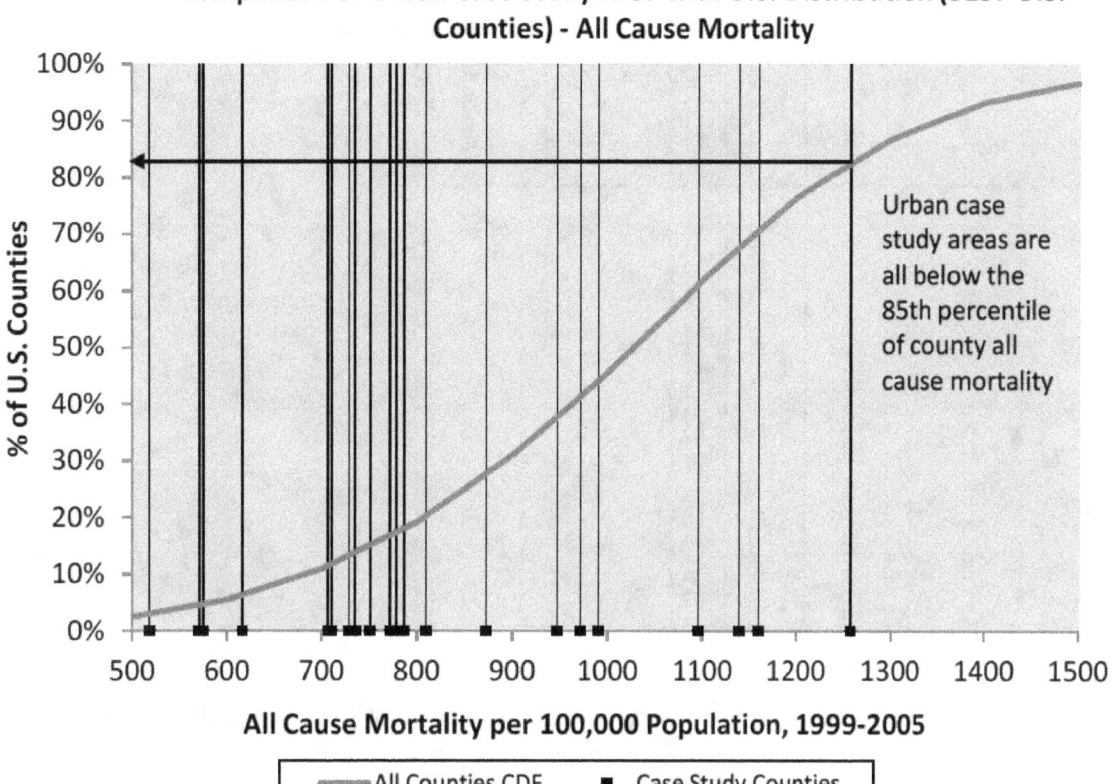

Comparison of Urban Case Study Area with U.S. Distribution (3137 U.S. Counties) - All Cause Mortality

Urban case study areas are all below the 85th percentile of county all cause mortality

% of U.S. Counties

All Cause Mortality per 100,000 Population, 1999-2005

All Counties CDF   ■ Case Study Counties

Figure A.9   Comparison of distributions for key elements of the risk equation: Baseline all-cause mortality

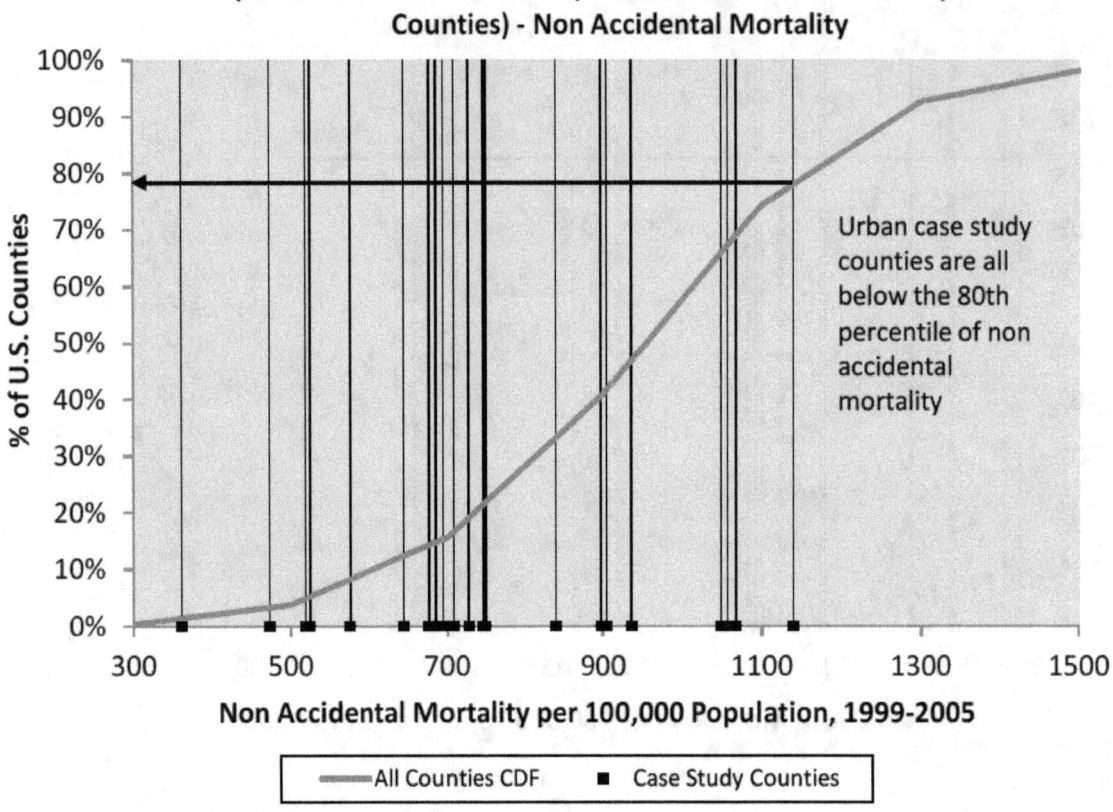

Figure A.10 **Comparison of distributions for key elements of the risk equation: Baseline non-accidental mortality**

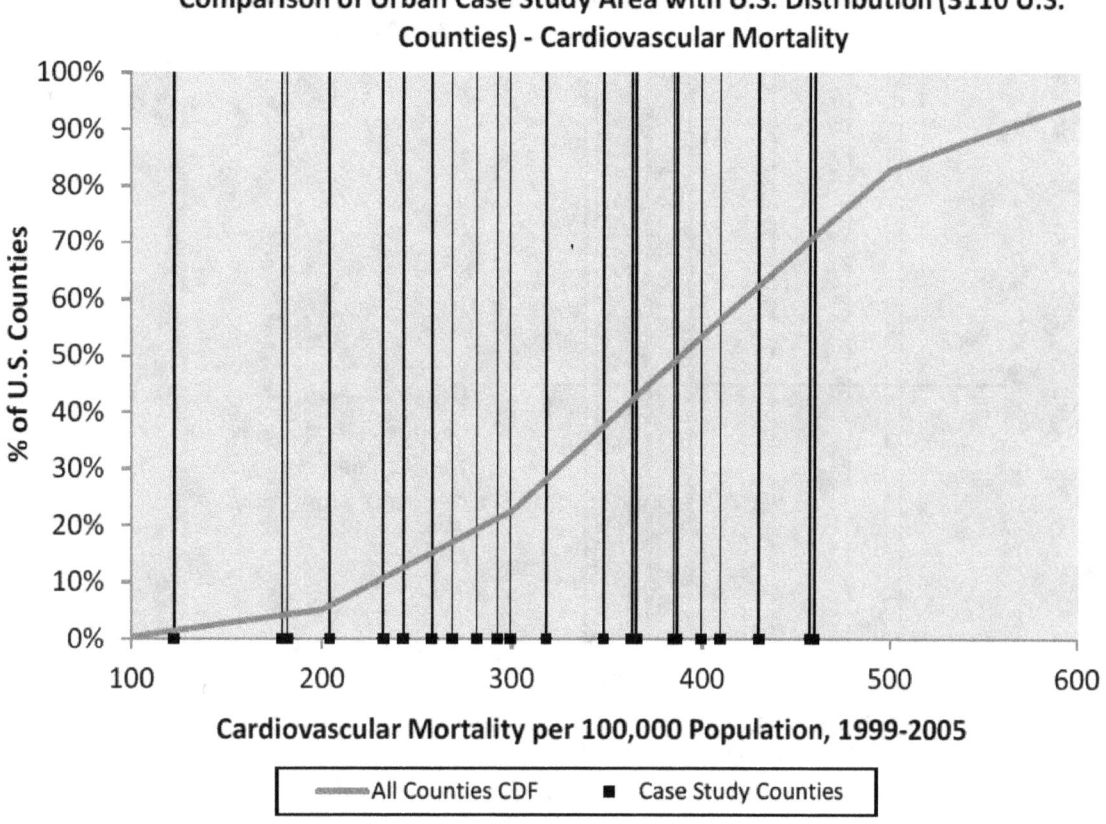

Figure A.11   Comparison of distributions for key elements of the risk equation: Baseline cardiovascular mortality

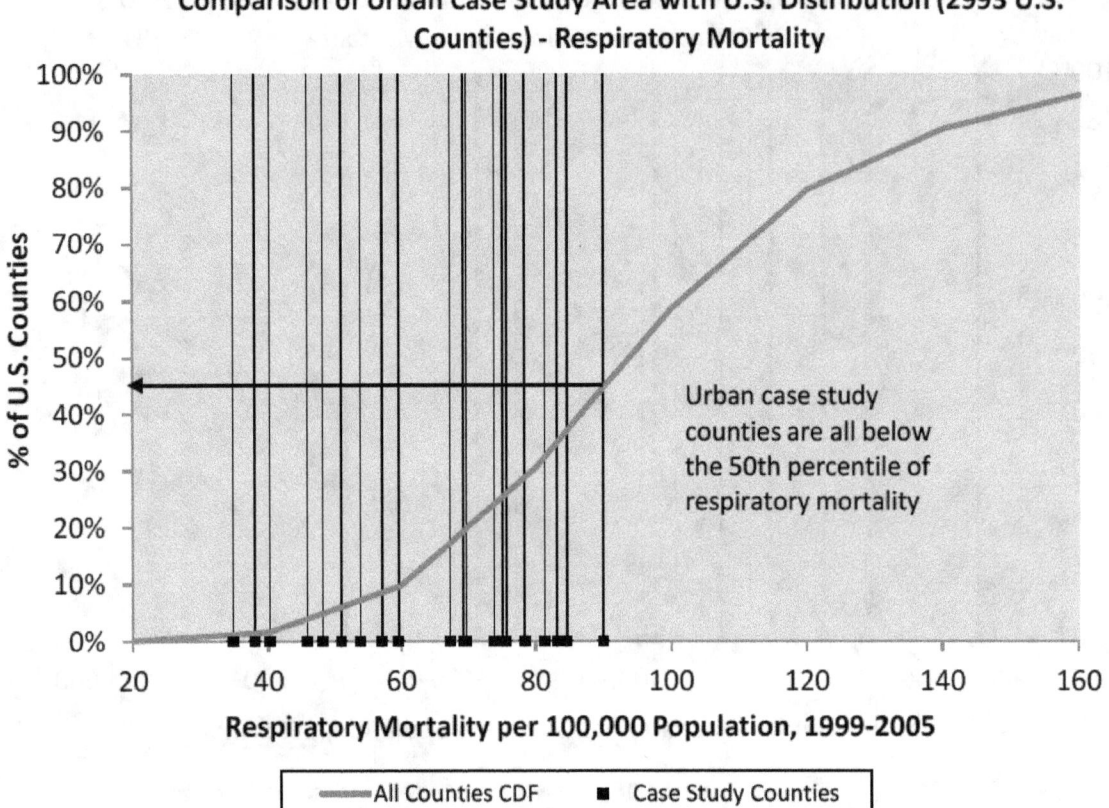

Figure A.12   Comparison of distributions for key elements of the risk equation: Baseline respiratory mortality

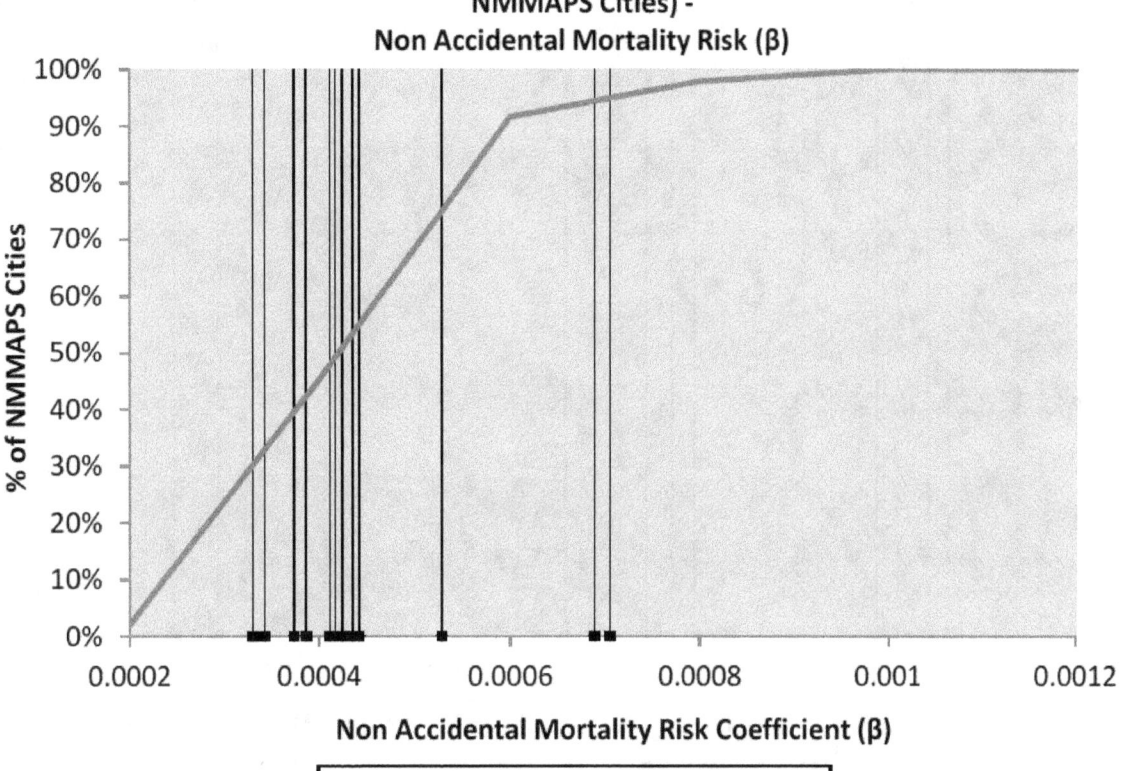

Figure A.13 Comparison of distributions for key elements of the risk equation: Non-accidental mortality risk coefficient from Bell et al. (2004)

13

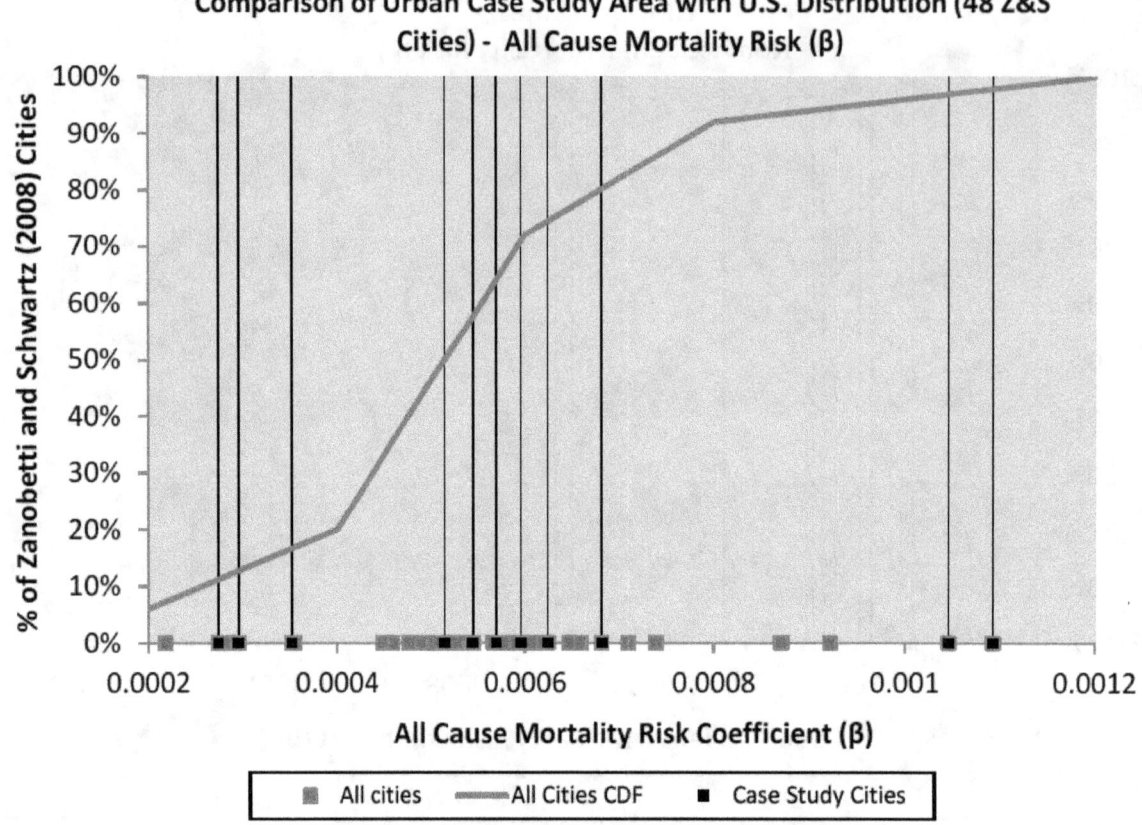

Figure A.14 Comparison of distributions for key elements of the risk equation: All-cause mortality risk coefficient from Zanobetti and Schwartz (2008)

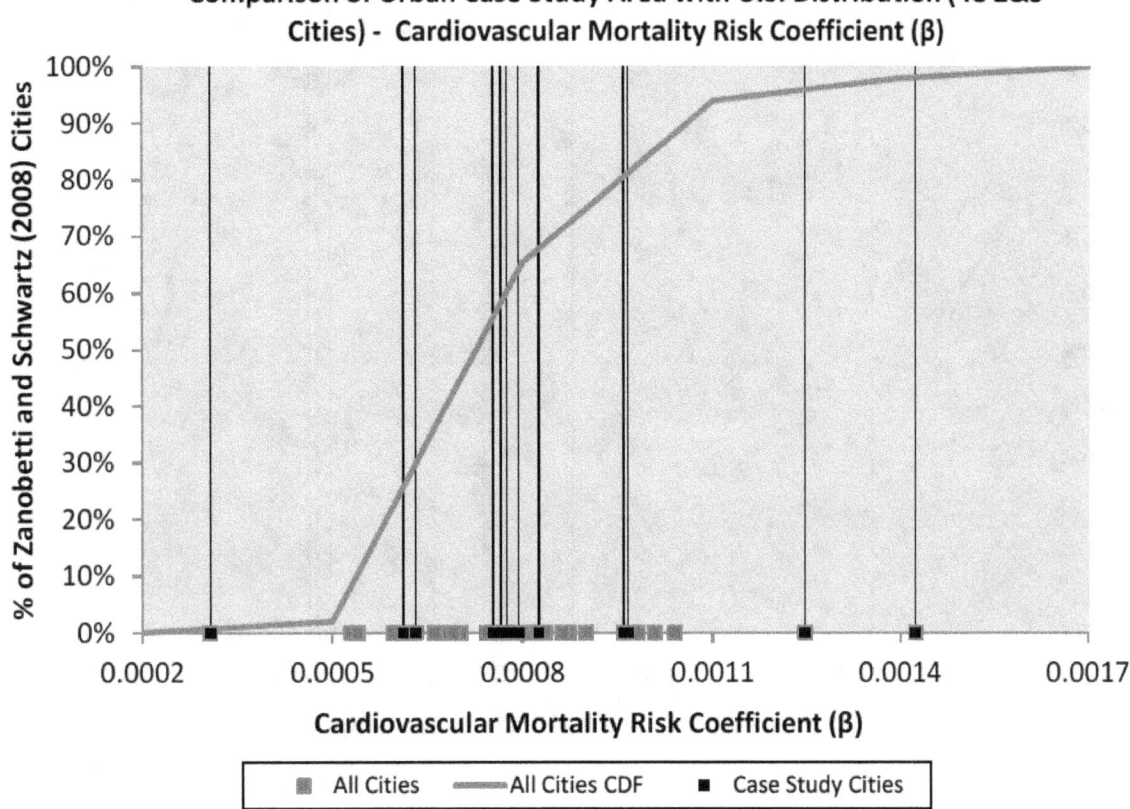

Figure A.15 Comparison of distributions for key elements of the risk equation: Cardiovascular mortality risk coefficient from Zanobetti and Schwartz (2008)

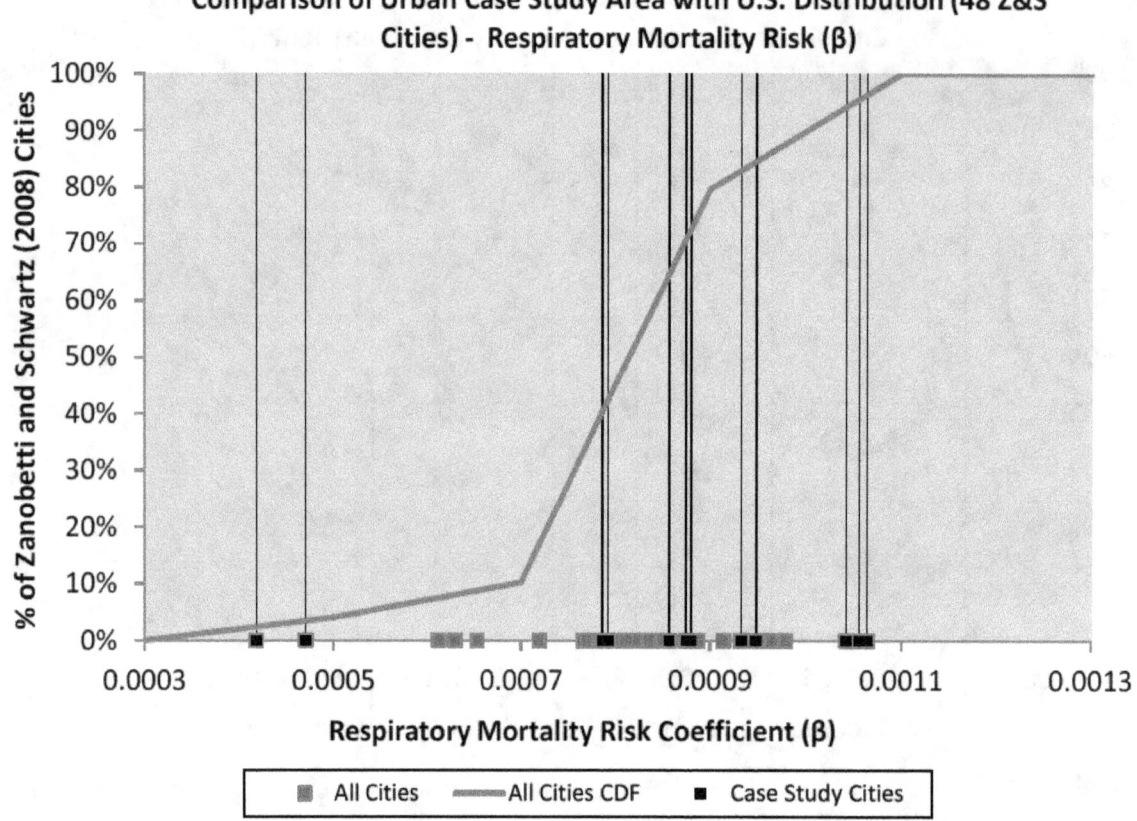

Figure A.16   Comparison of distributions for key elements of the risk equation:
Respiratory mortality risk coefficient from Zanobetti and Schwartz (2008)

Variables Expected to Influence the Relative Risk from Ozone

Demographic Variables

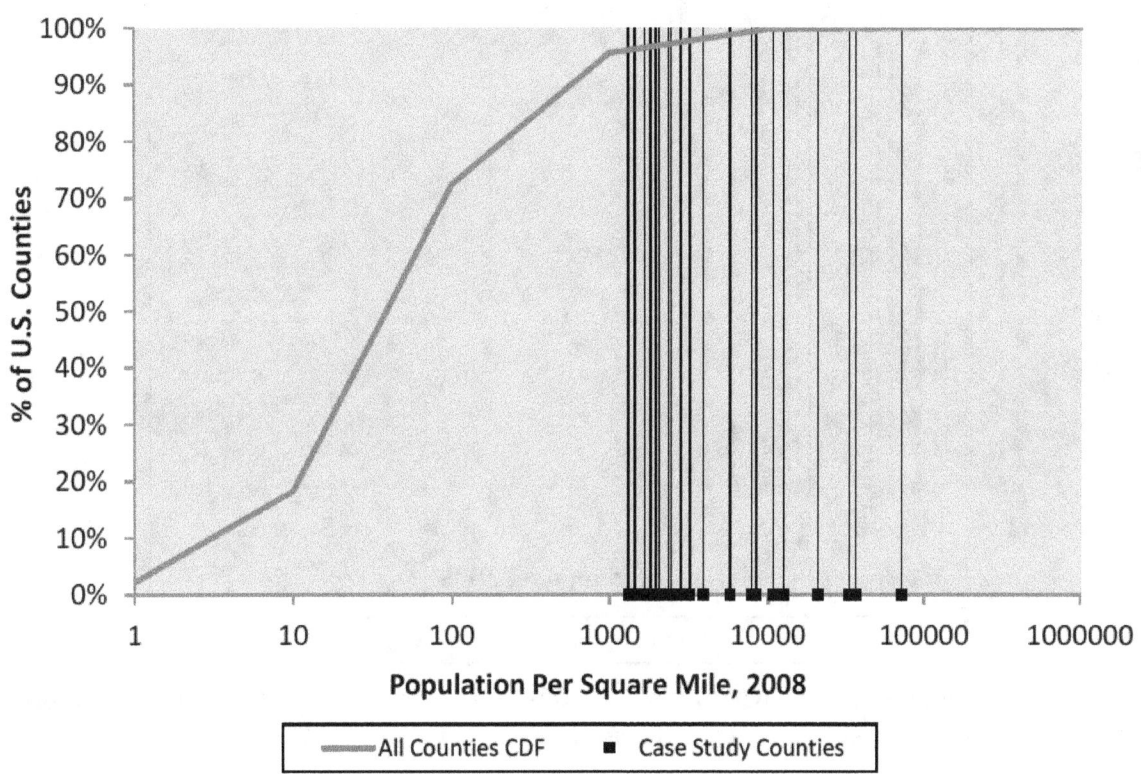

Figure A.17   Comparison of distributions for selected variables expected to influence the relative risk from ozone: Population density

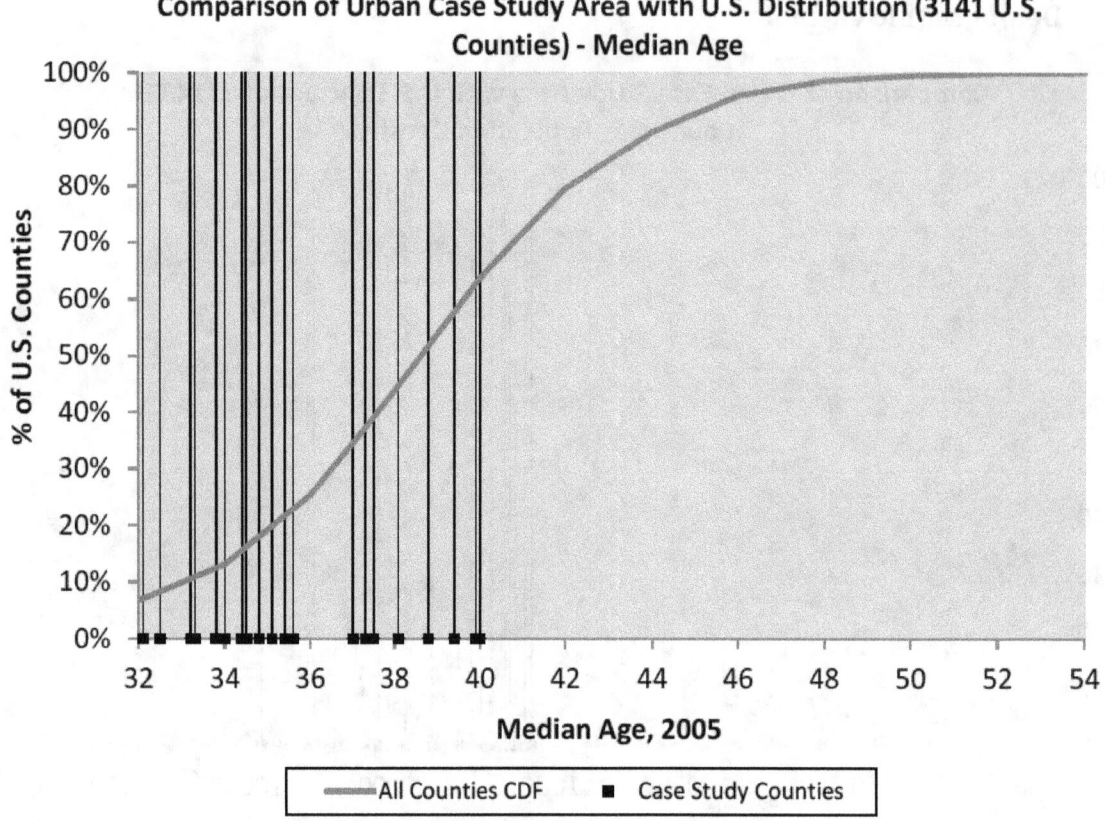

Figure A.18 Comparison of distributions for selected variables expected to influence the relative risk from ozone: Median age

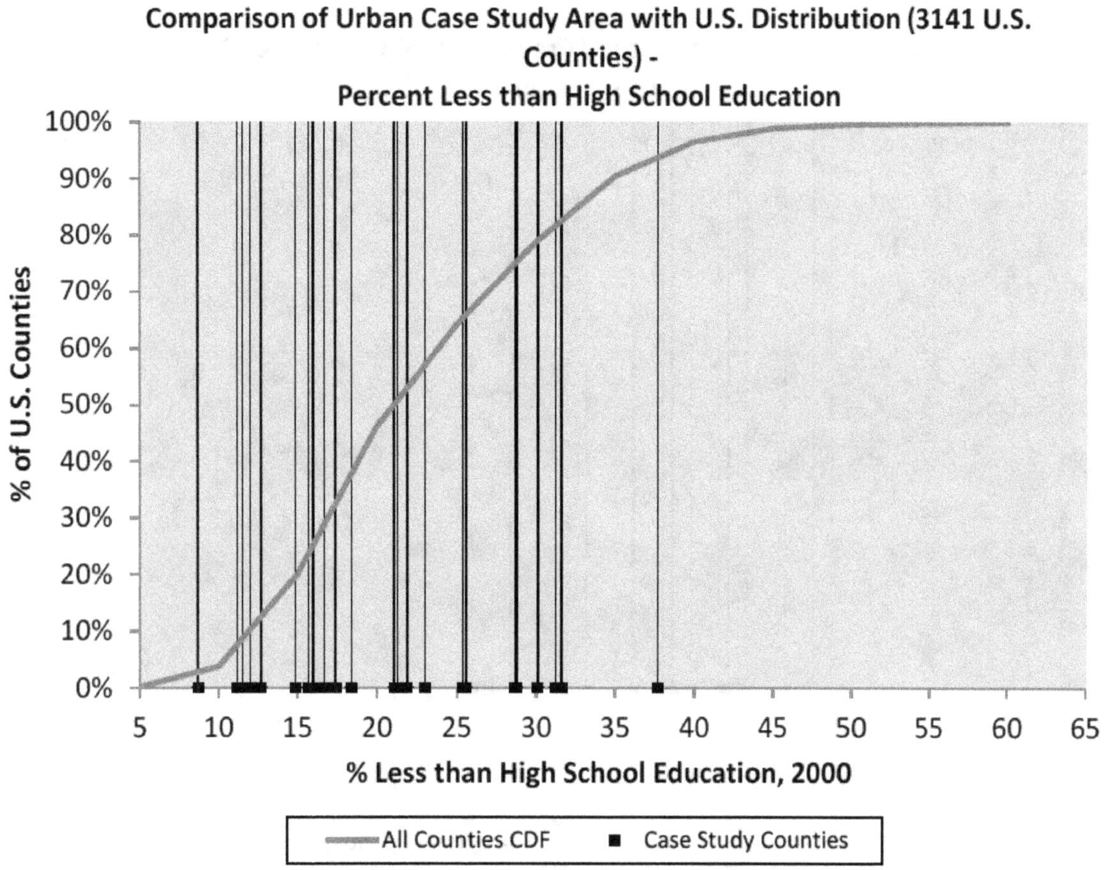

Figure A.19   Comparison of distributions for selected variables expected to influence the relative risk from ozone: Percent less than high school education

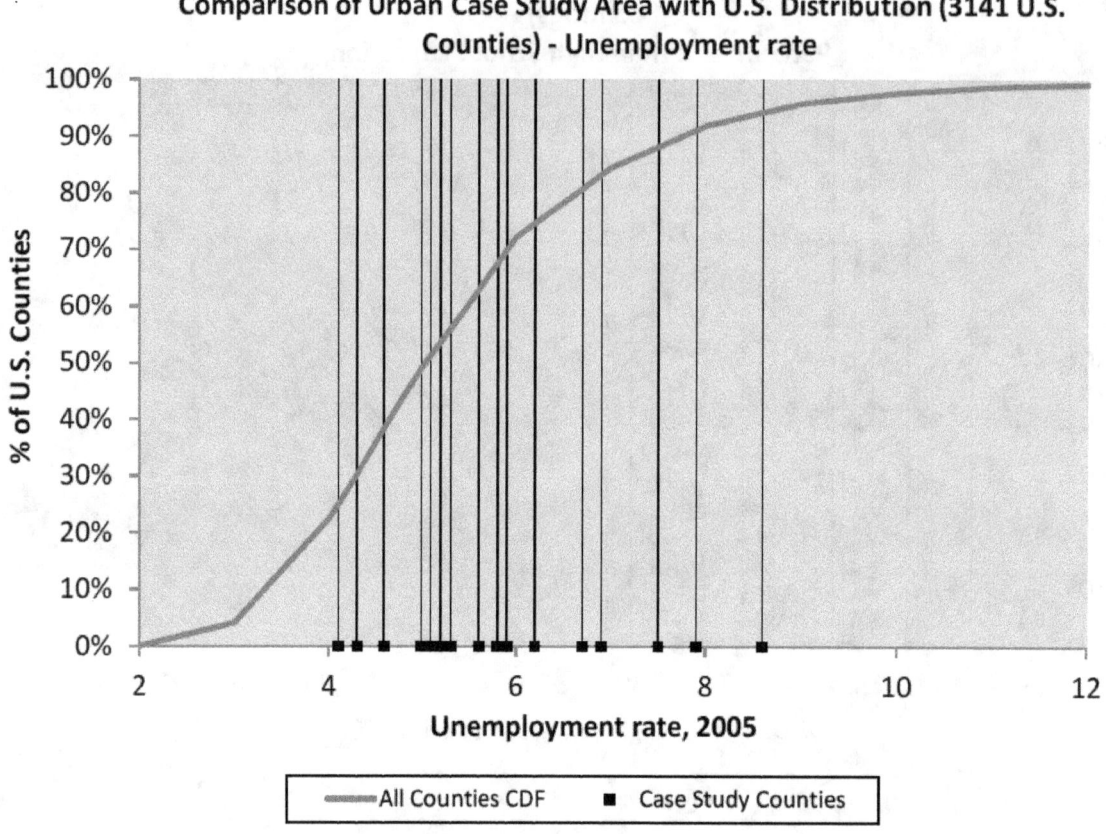

Figure A.20 Comparison of distributions for selected variables expected to influence the relative risk from ozone: Unemployment rate

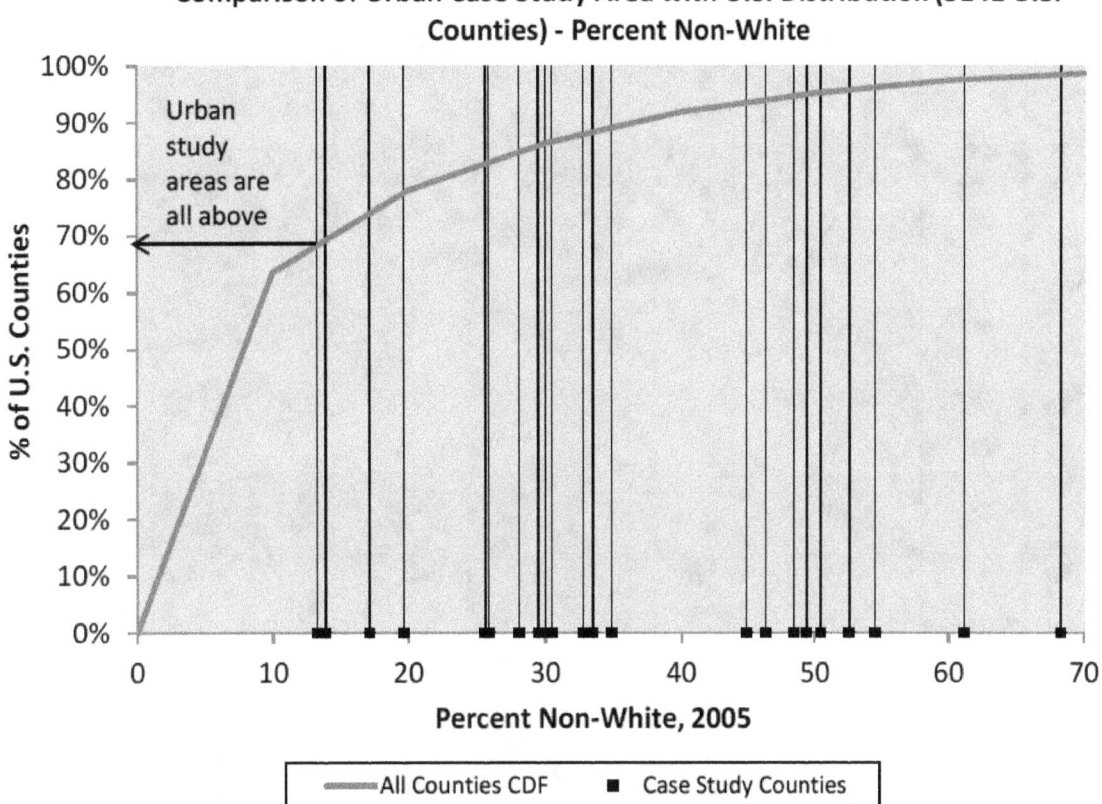

Figure A.21   Comparison of distributions for selected variables expected to influence the relative risk from ozone: Percent non-white

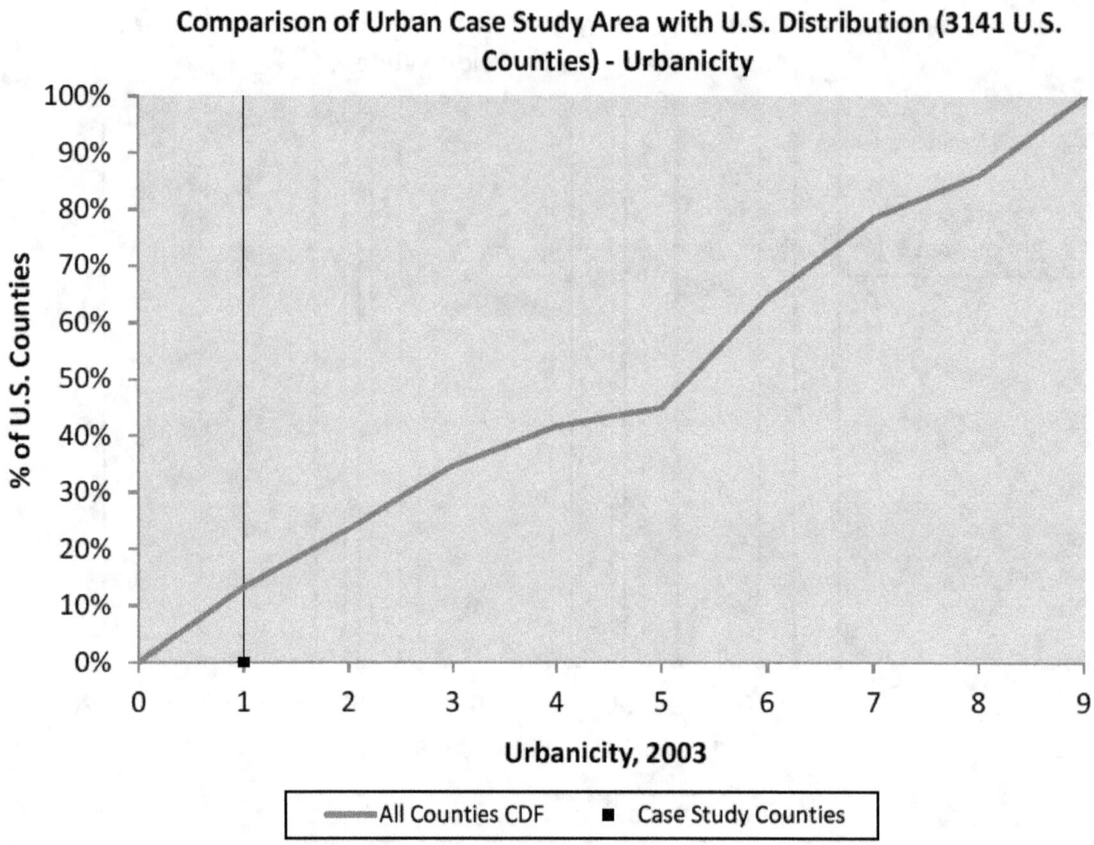

Figure A.22 Comparison of distributions for selected variables expected to influence the relative risk from ozone: Urbanicity

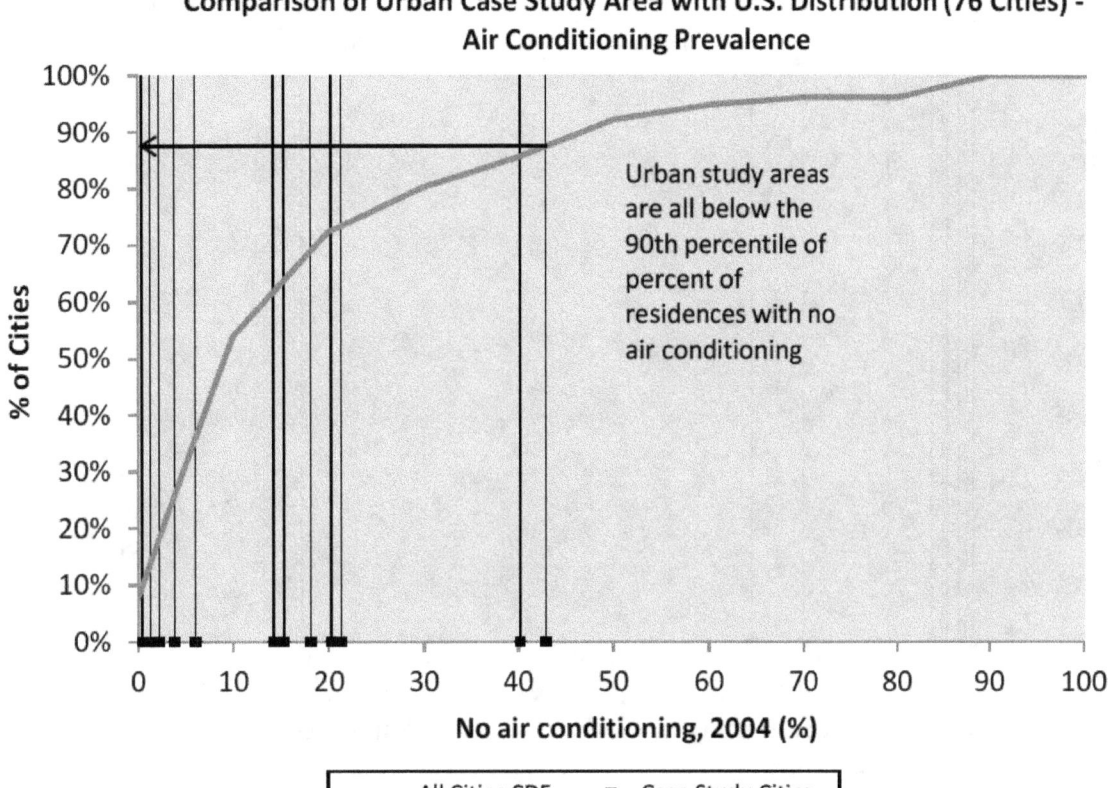

Figure A.23 **Comparison of distributions for selected variables expected to influence the relative risk from ozone: Air conditioning prevalence**

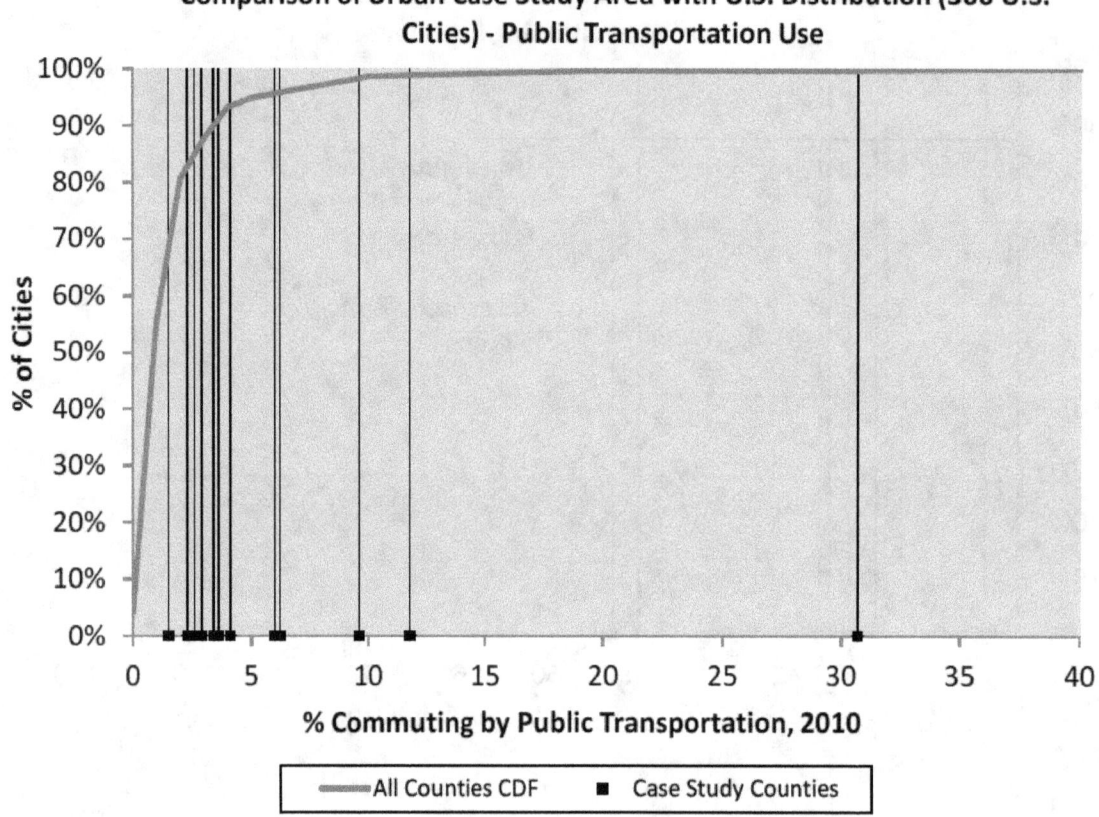

Figure A.24   Comparison of distributions for selected variables expected to influence the relative risk from ozone: Percent commuting by public transportation

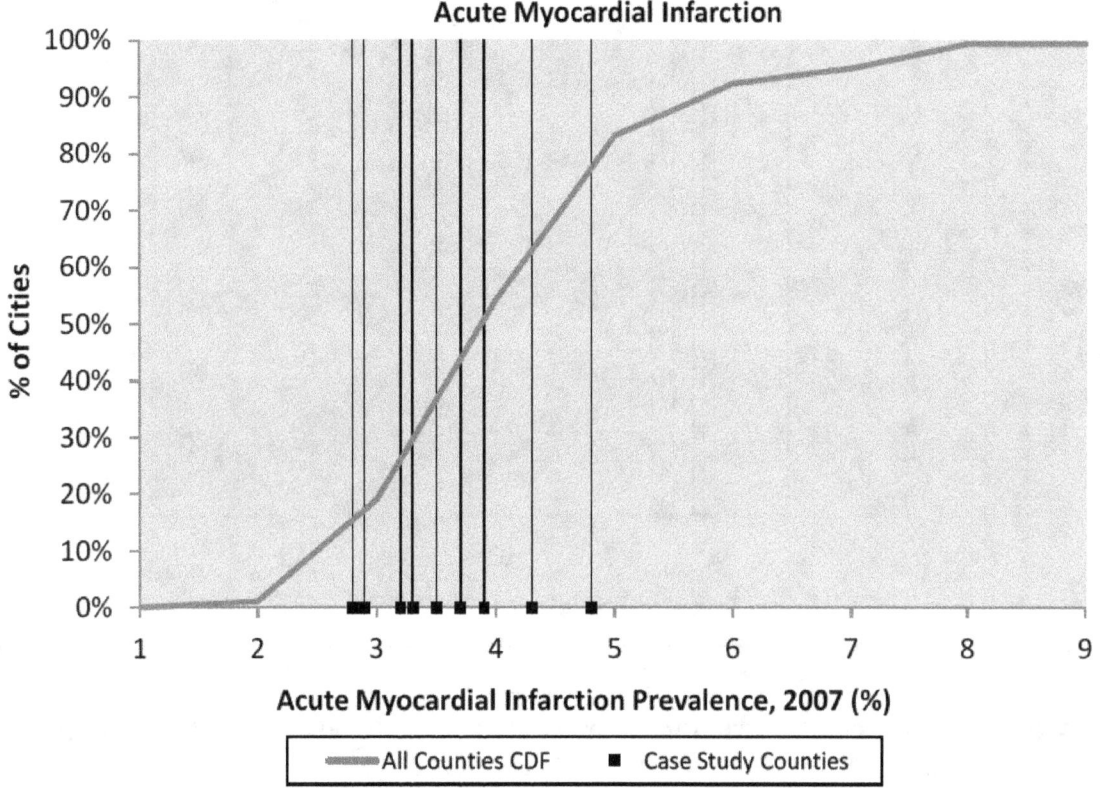

Figure A.25   Comparison of distributions for selected variables expected to influence the relative risk from ozone: Acute myocardial infarction prevalence

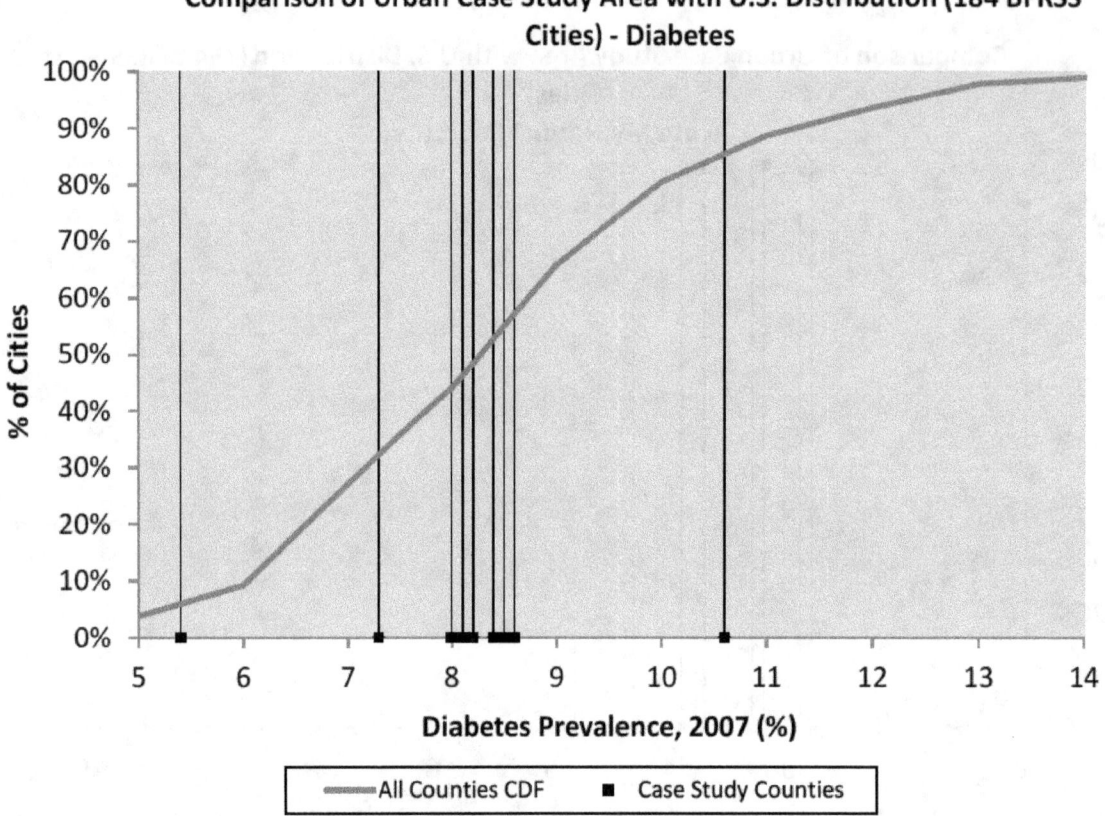

Figure A.26 **Comparison of distributions for selected variables expected to influence the relative risk from ozone: Diabetes prevalence**

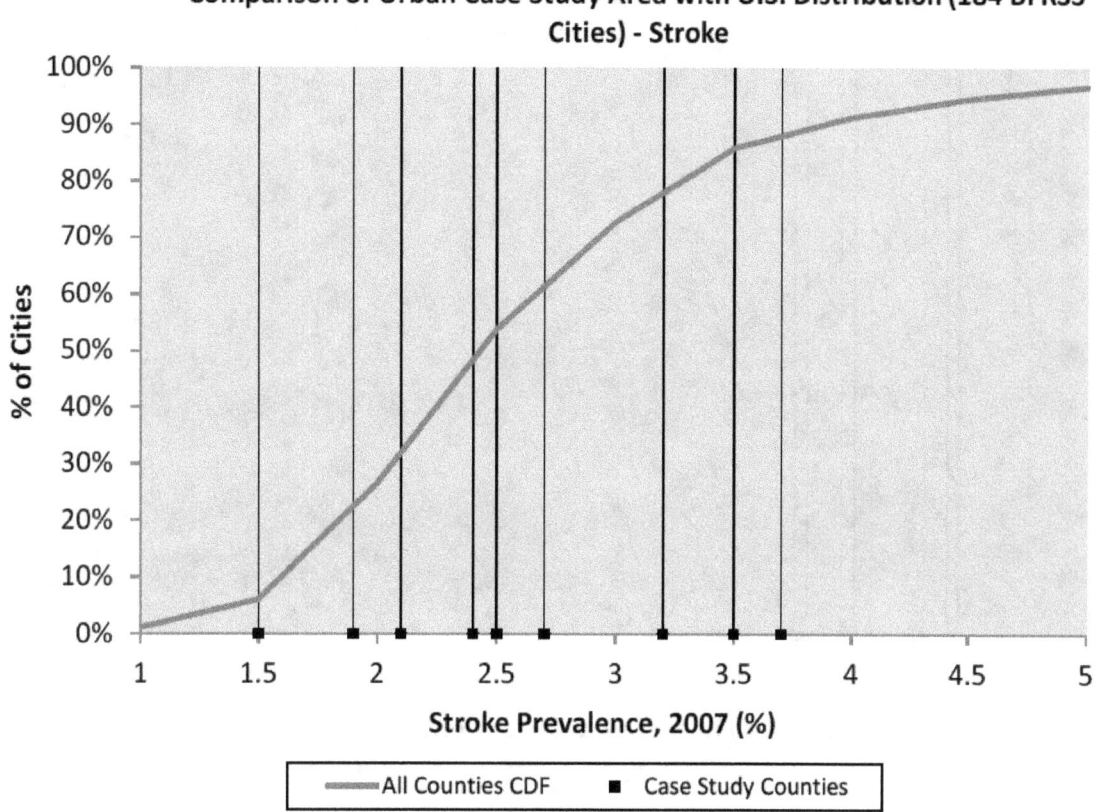

Figure A.27 Comparison of distributions for selected variables expected to influence the relative risk from ozone: Stroke prevalence

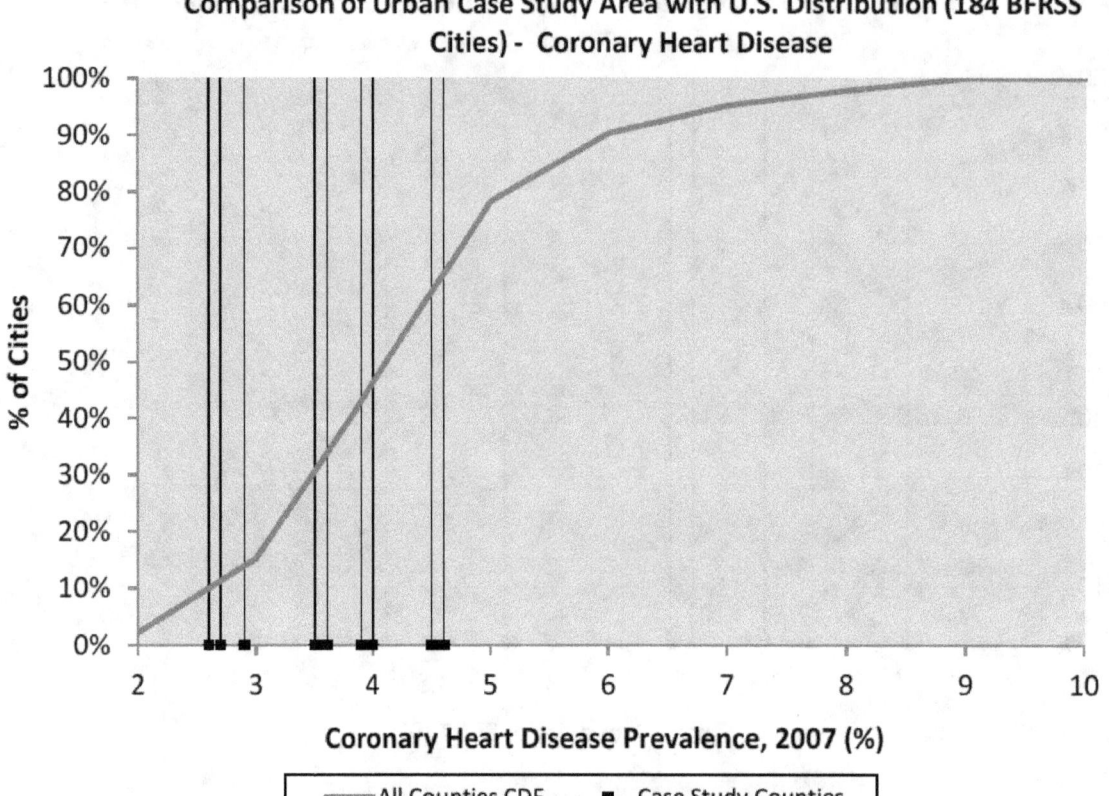

Figure A.28   Comparison of distributions for selected variables expected to influence the relative risk from ozone: Coronary heart disease prevalence

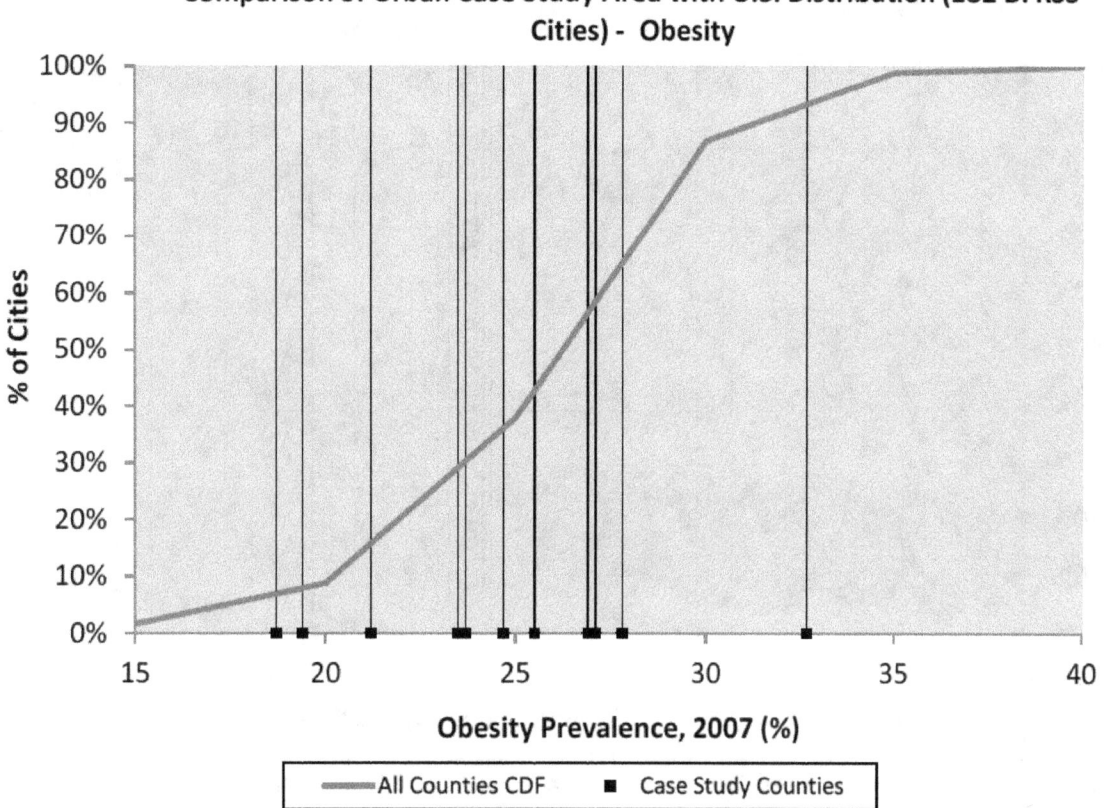

Figure A.29   Comparison of distributions for selected variables expected to influence the relative risk from ozone: Obesity prevalence

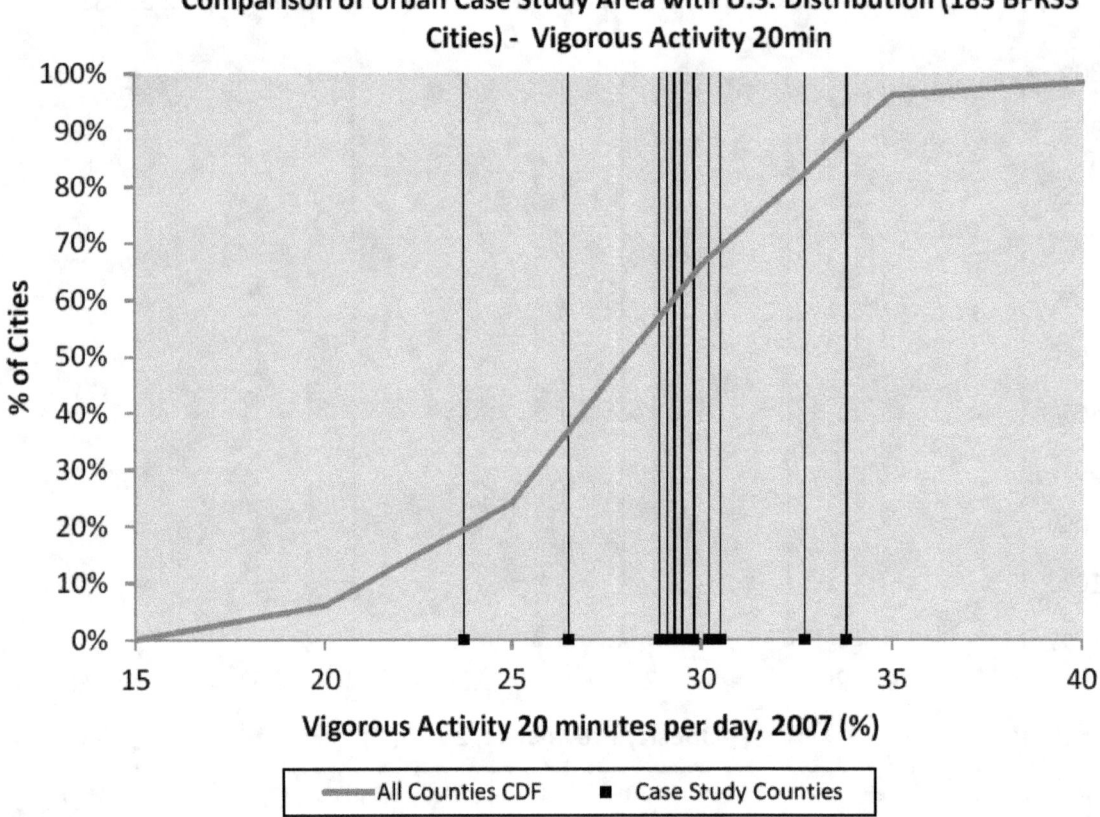

Comparison of Urban Case Study Area with U.S. Distribution (183 BFRSS Cities) - Vigorous Activity 20min

Figure A.30 Comparison of distributions for selected variables expected to influence the relative risk from ozone: Vigorous activity at least 20 minutes per day

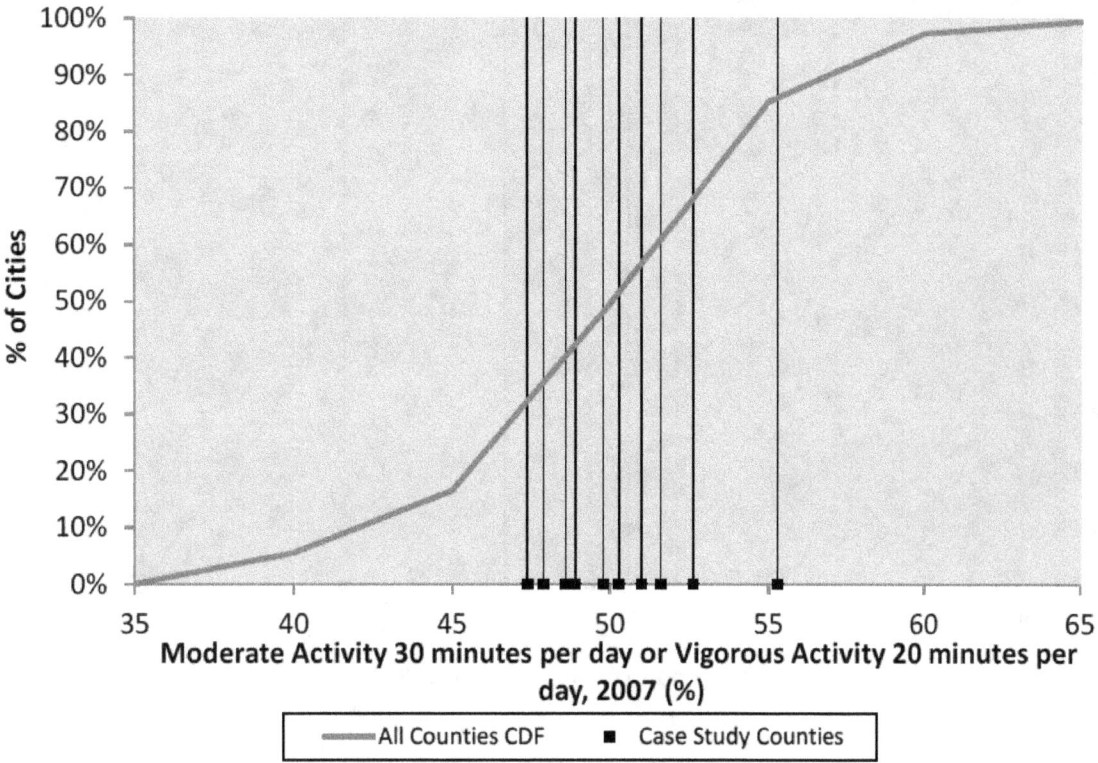

Figure A.31 Comparison of distributions for selected variables expected to influence the relative risk from ozone: Moderate activity at least 30 minutes per day or vigorous activity at least 20 minutes per day

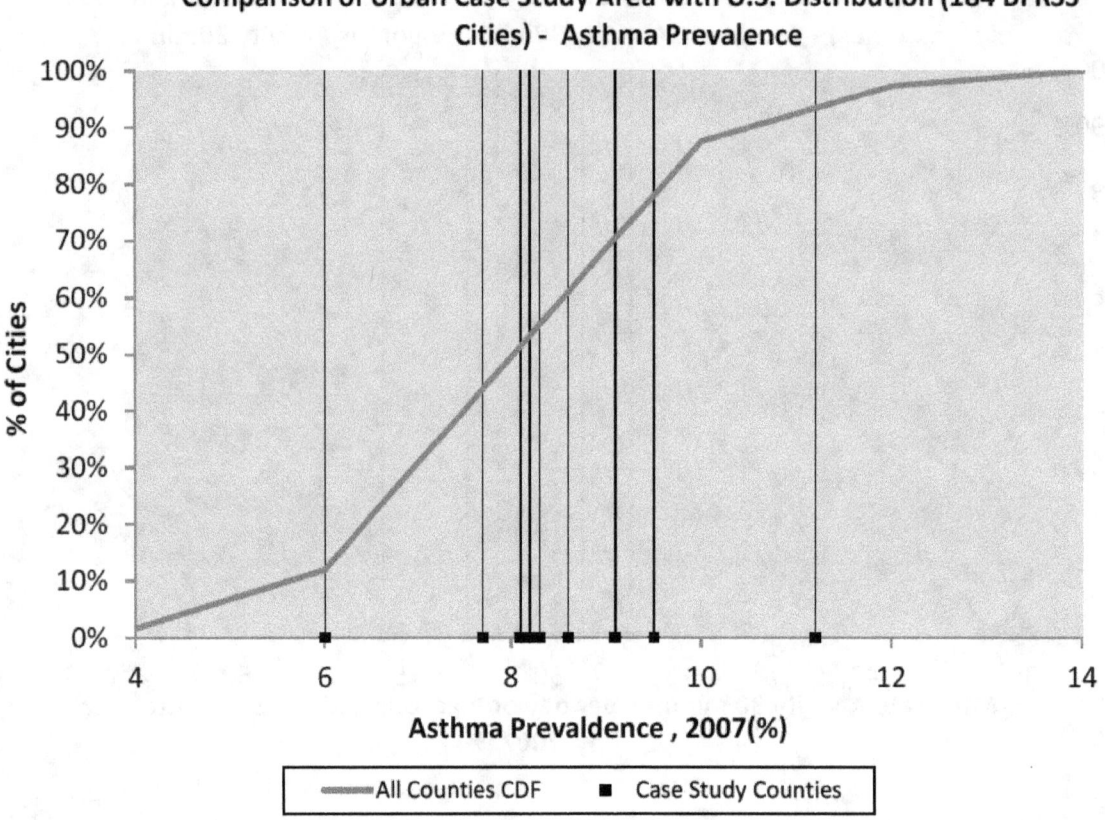

Figure A.32   Comparison of distributions for selected variables expected to influence the relative risk from ozone: Asthma prevalence

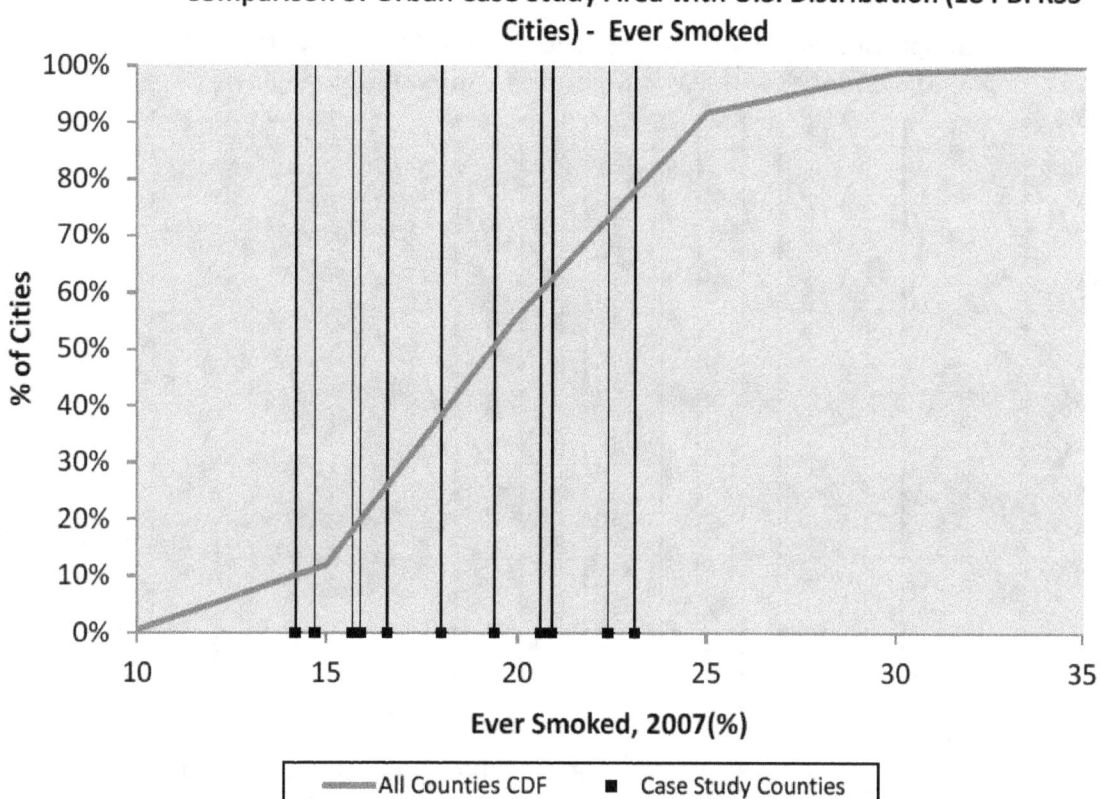

Comparison of Urban Case Study Area with U.S. Distribution (184 BFRSS Cities) - Ever Smoked

Figure A.33 Comparison of distributions for selected variables expected to influence the relative risk from ozone: Smoking prevalence

Air Quality and Climate Variables

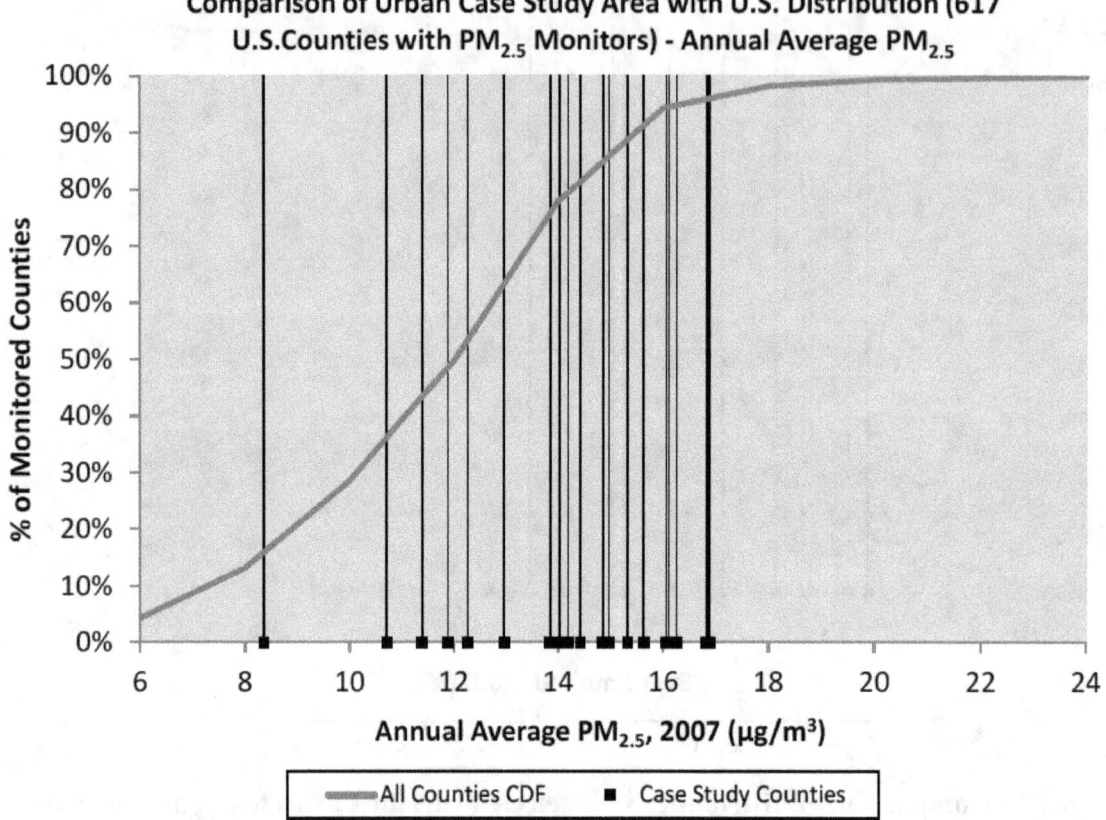

Figure A.34   Comparison of distributions for selected variables expected to influence the relative risk from ozone: Annual average PM$_{2.5}$ concentration

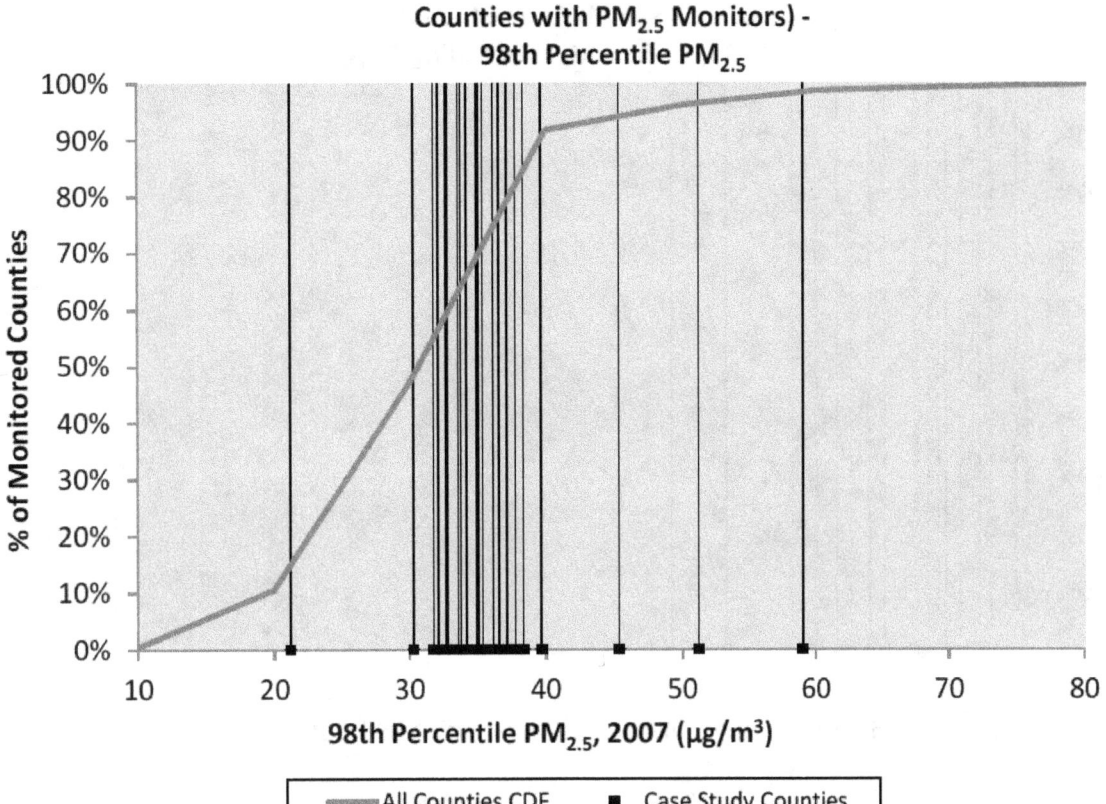

Figure A.35   Comparison of distributions for selected variables expected to influence the relative risk from ozone: 98th percentile PM2.5 concentration

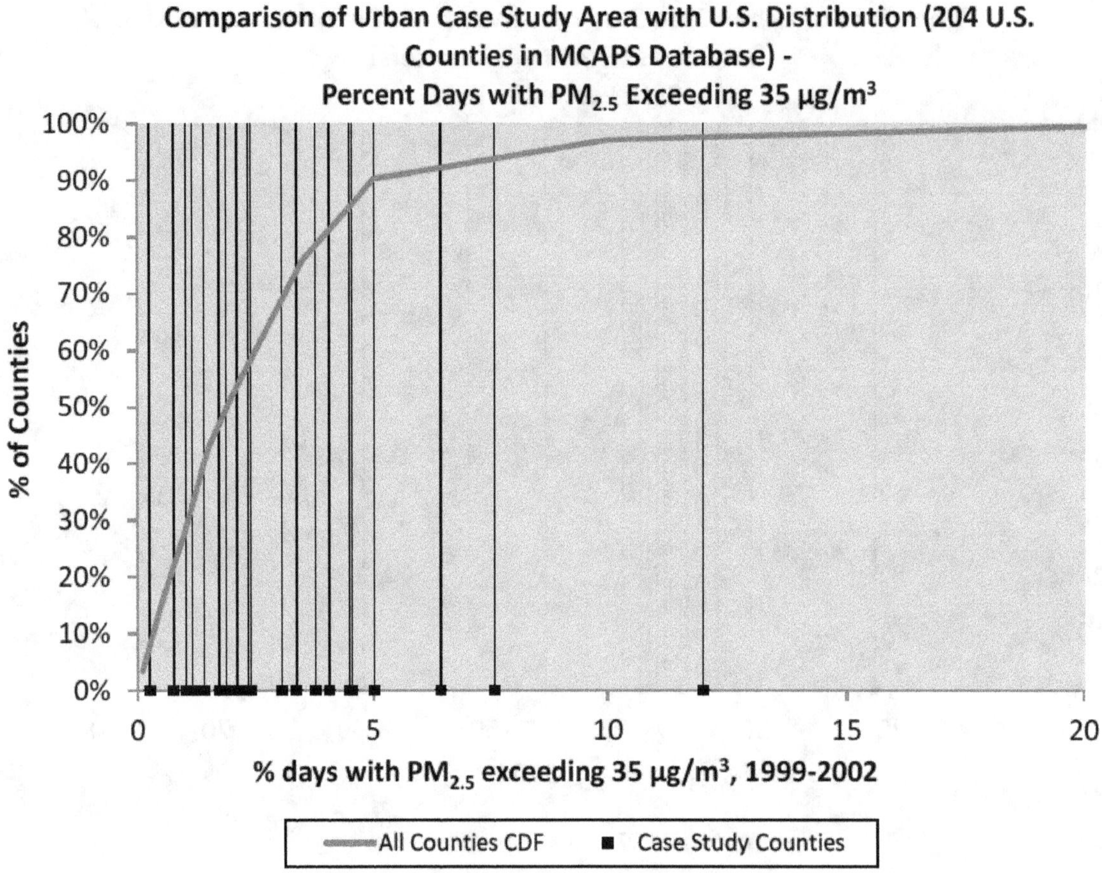

Comparison of Urban Case Study Area with U.S. Distribution (204 U.S. Counties in MCAPS Database) - Percent Days with PM$_{2.5}$ Exceeding 35 µg/m^3

Figure A.36 Comparison of distributions for selected variables expected to influence the relative risk from ozone: Percent of days with PM$_{2.5}$ exceeding 35 µg/m^3

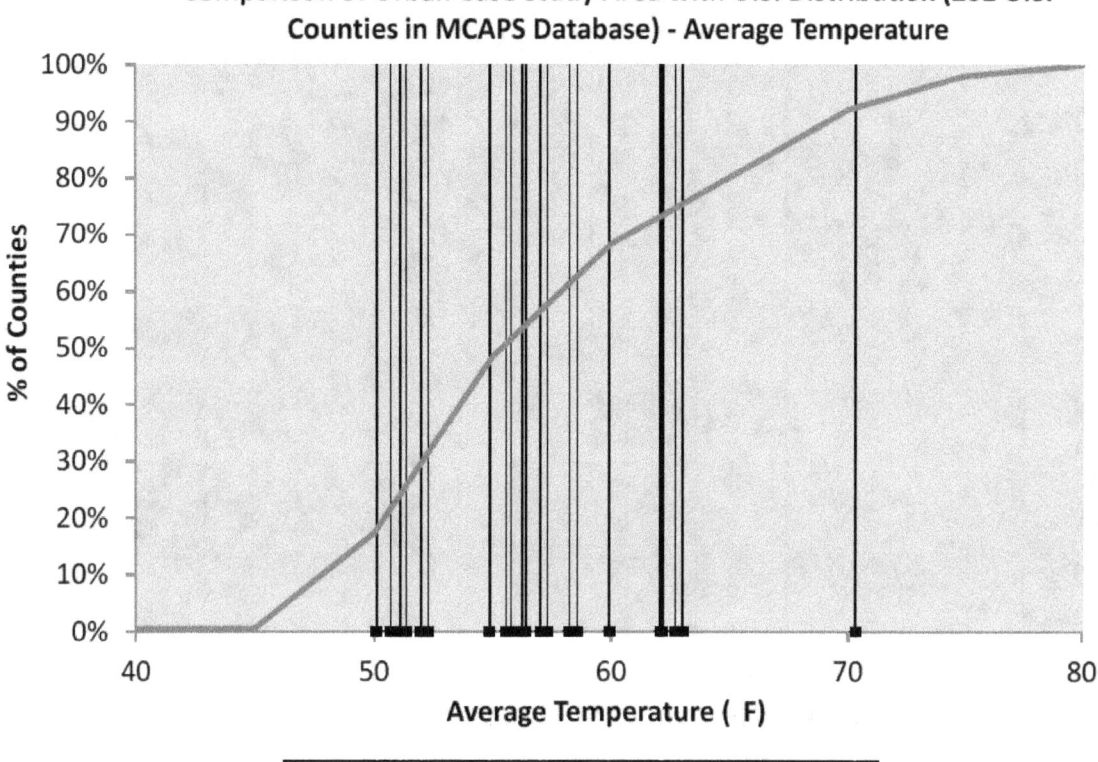

Figure A.37 Comparison of distributions for selected variables expected to influence the relative risk from ozone: Average temperature

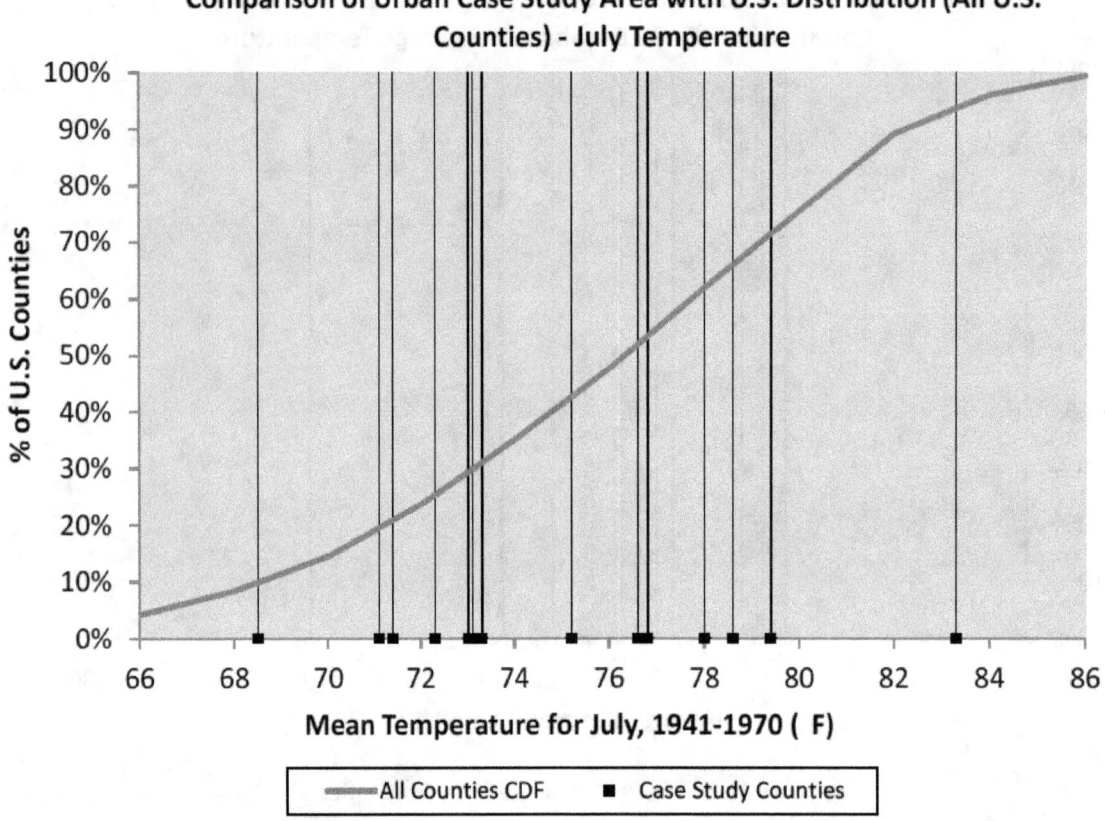

Figure A.38   Comparison of distributions for selected variables expected to influence the relative risk from ozone: July temperature

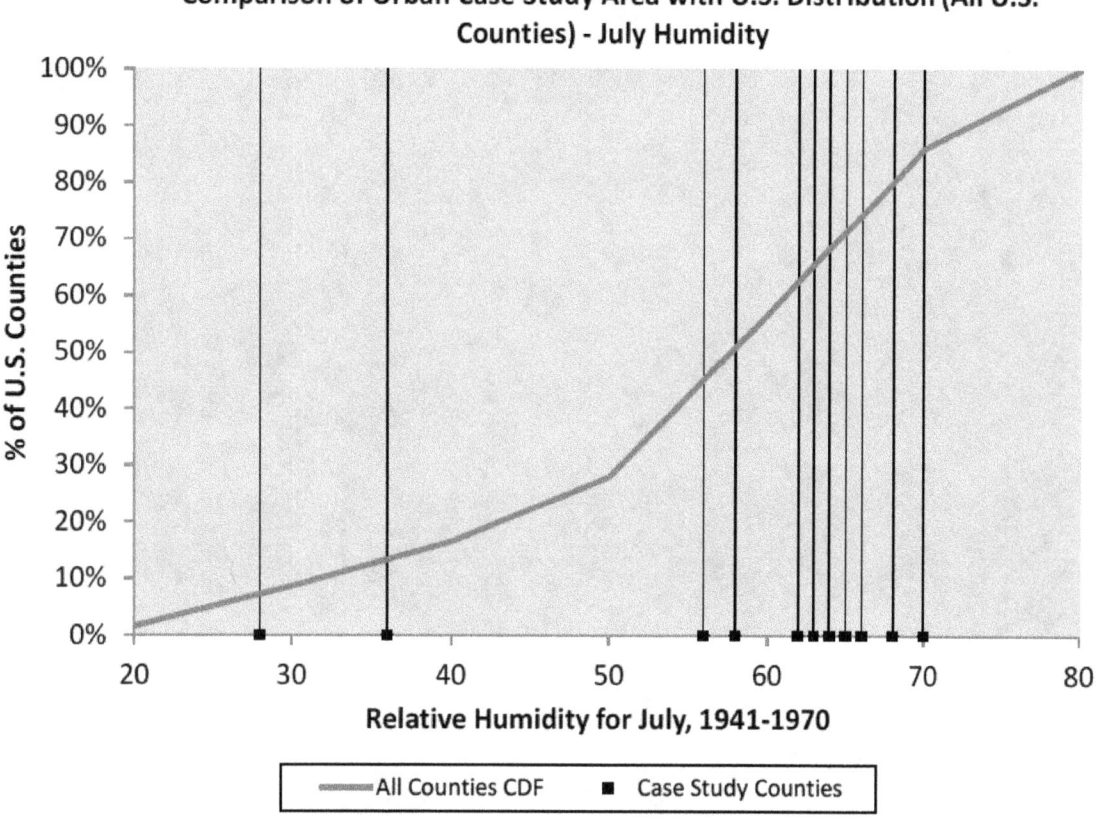

Figure A.39   Comparison of distributions for selected variables expected to influence the relative risk from ozone: Relative humidity

8-C. National Representativeness of Ozone Response to Emissions Changes

Table of Contents

TABLE OF FIGURES

Table of Tables

This appendix provides additional plots and information to support the analysis provided in section 8.2.3 of the main text of the health REA.

1. AMBIENT TRENDS OVER A PERIOD OF NATIONALLY DECREASING NOX EMISSIONS

1.1 NATIONWIDE MAPS SHOWING ABSOLUTE CHANGES IN OZONE BETWEEN 2001-2003 AND 2008-2010

In Chapter 8 we provided maps of US ozone monitors showing absolute changes in ozone percentiles between a 3-year period before many of the nationwide NOx reductions took place (2001-2003) and a period after many of these reductions took place (2008-2010). Here we provide a full set of maps which includes not only the behavior of the 50th and 95th percentiles but also 5th, 25th, and 75th percentiles for three different groupings of months: short summer season (June-August), longer warm season (April-October), and all year. These plots further support the general trends that were noted in chapter 8: ozone increases occurred more in cooler months than warmer months, ozone increases occurred more at the lower end of the distribution that the upper end of the distribution, and ozone increases were more likely to occur in urban core area than at locations further from the city centers. The plots of 95th percentile ozone changes show that high ozone days have decreased across the country at all times of year. The June-August plots show that mid-range ozone has also decreased at most locations during the warmest time of year when ozone levels are highest.

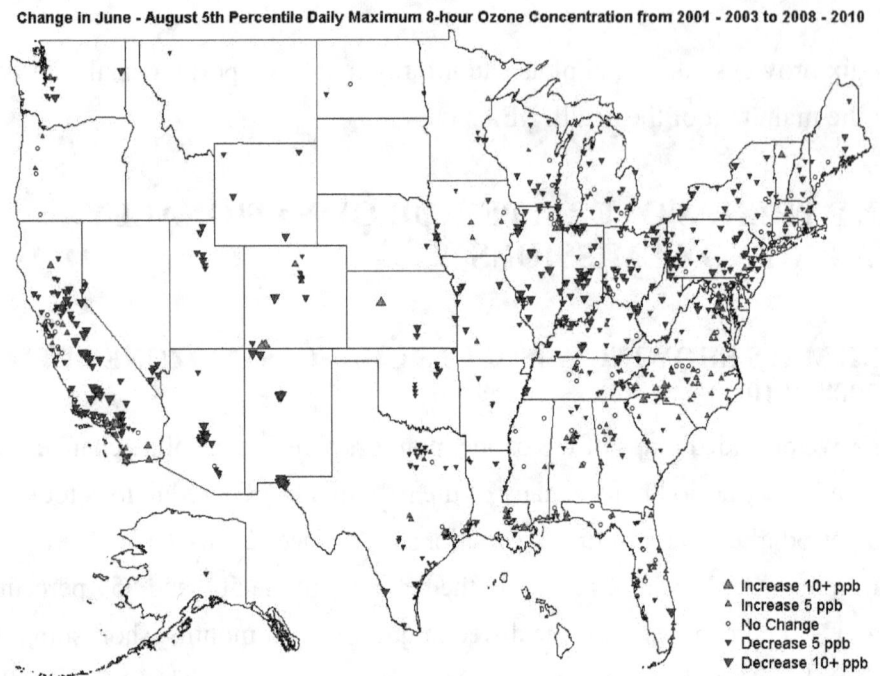

Figure 1: Change in 5th percentile June-August summer season daily 8-hour maximum ozone concentrations between 2001-2003 and 2008-2010.

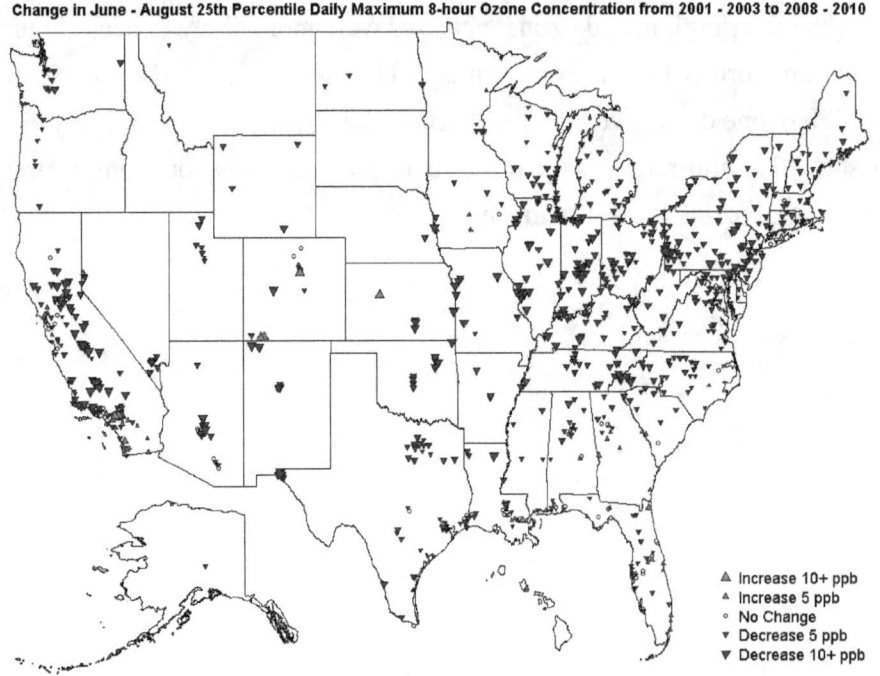

Figure 2: Change in 25th percentile June-August summer season daily 8-hour maximum ozone concentrations between 2001-2003 and 2008-2010.

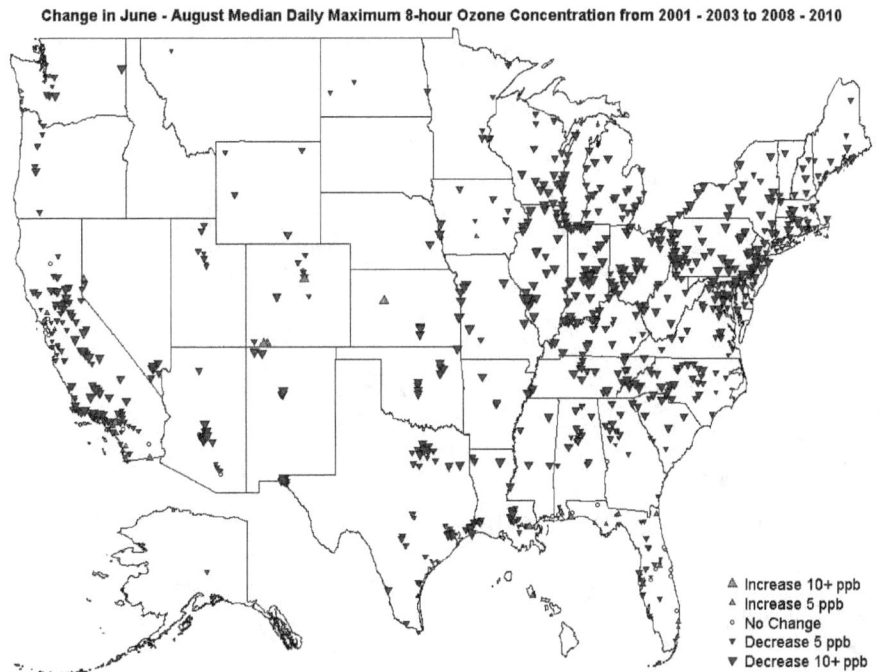

Figure 3: Change in 50th percentile June-August summer season daily 8-hour maximum ozone concentrations between 2001-2003 and 2008-2010.

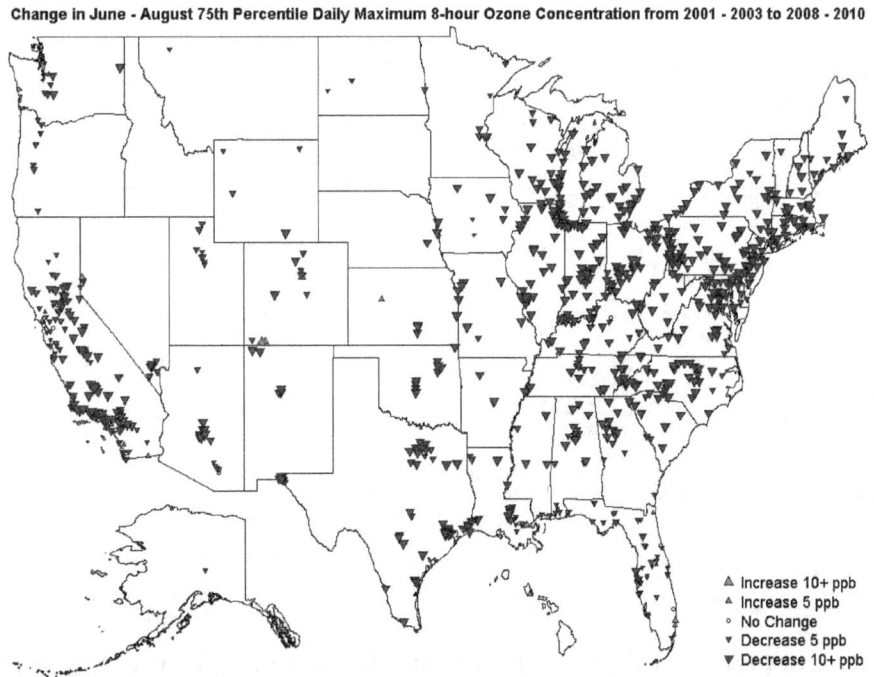

Figure 4: Change in 75th percentile June-August summer season daily 8-hour maximum ozone concentrations between 2001-2003 and 2008-2010.

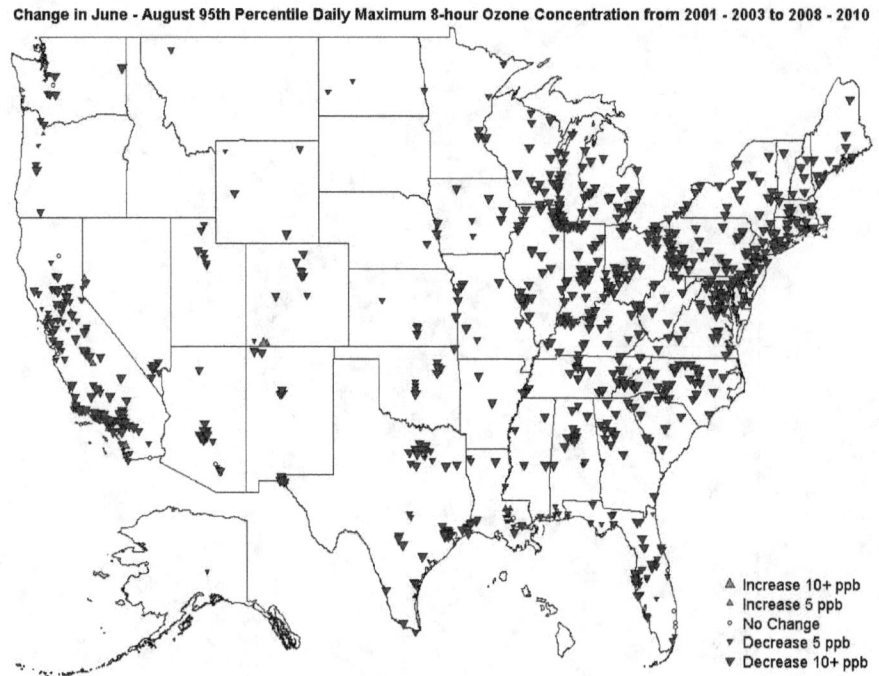

Figure 5: Change in 95th percentile June-August summer season daily 8-hour maximum ozone concentrations between 2001-2003 and 2008-2010.

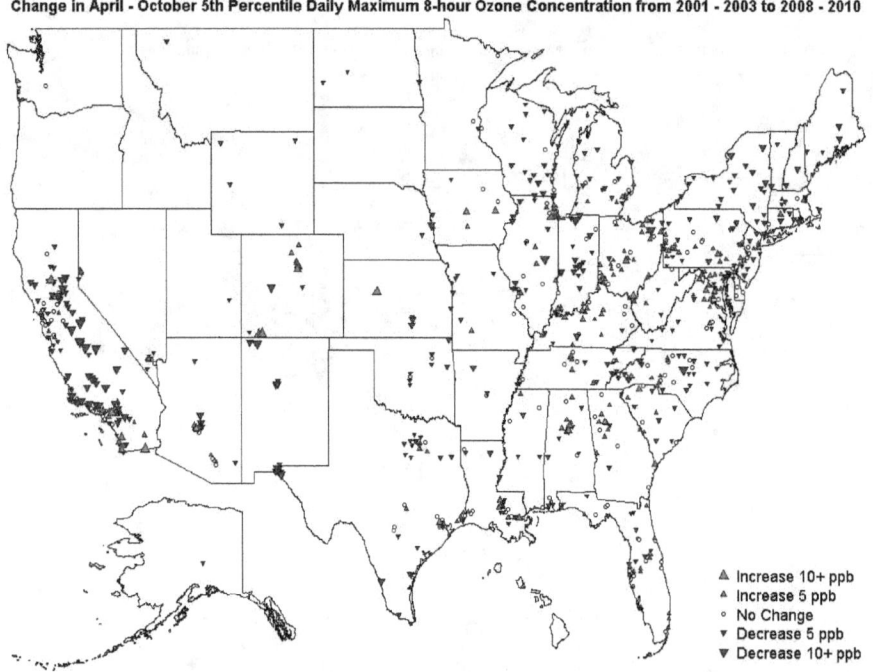

Figure 6: Change in 5th percentile April-October summer season daily 8-hour maximum ozone concentrations between 2001-2003 and 2008-2010.

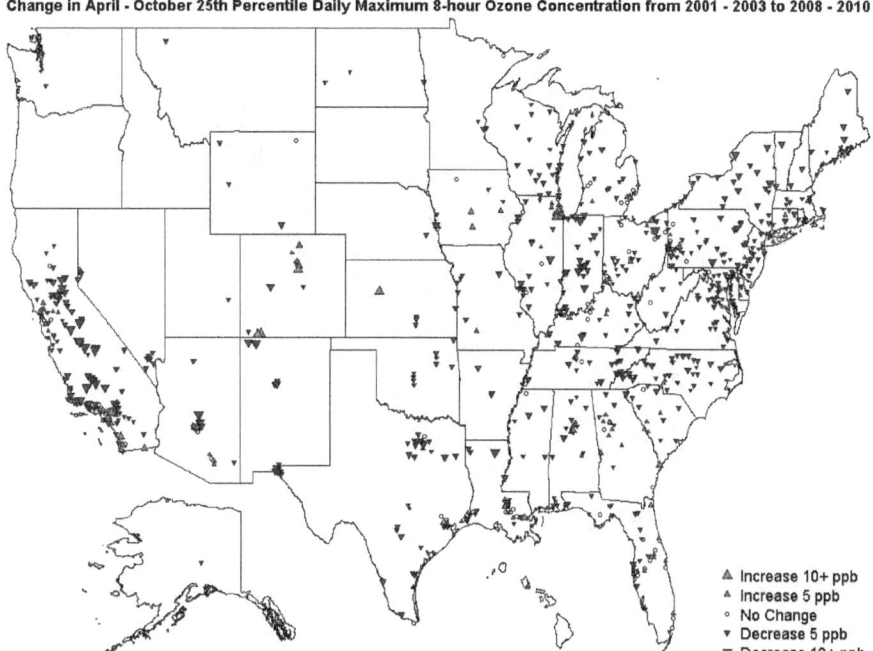

Change in April - October 25th Percentile Daily Maximum 8-hour Ozone Concentration from 2001 - 2003 to 2008 - 2010

△ Increase 10+ ppb
▲ Increase 5 ppb
○ No Change
▼ Decrease 5 ppb
▼ Decrease 10+ ppb

Figure 7: Change in 25th percentile April-October summer season daily 8-hour maximum ozone concentrations between 2001-2003 and 2008-2010.

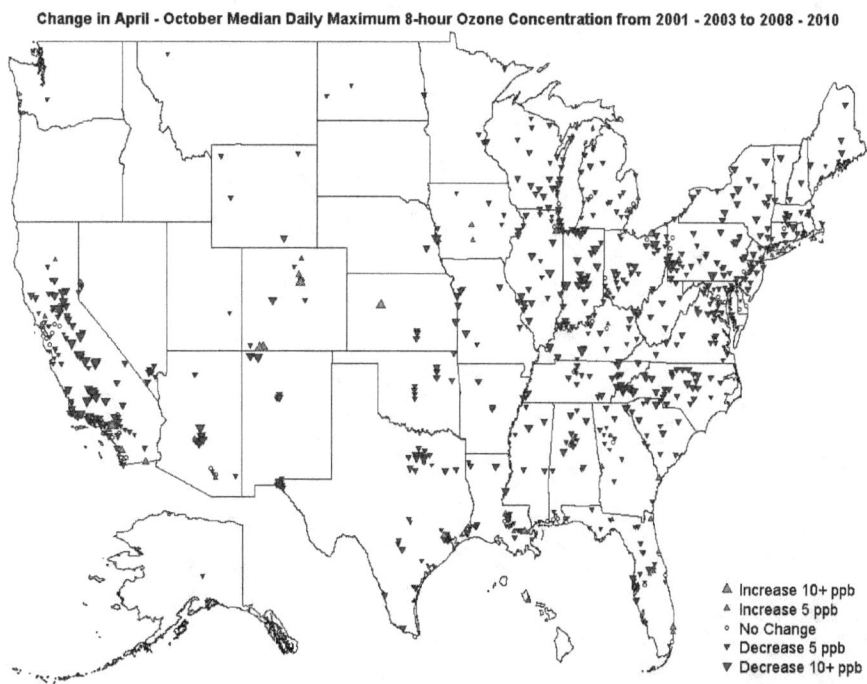

Change in April - October Median Daily Maximum 8-hour Ozone Concentration from 2001 - 2003 to 2008 - 2010

△ Increase 10+ ppb
▲ Increase 5 ppb
○ No Change
▼ Decrease 5 ppb
▼ Decrease 10+ ppb

Figure 8: Change in 50th percentile April-October summer season daily 8-hour maximum ozone concentrations between 2001-2003 and 2008-2010.

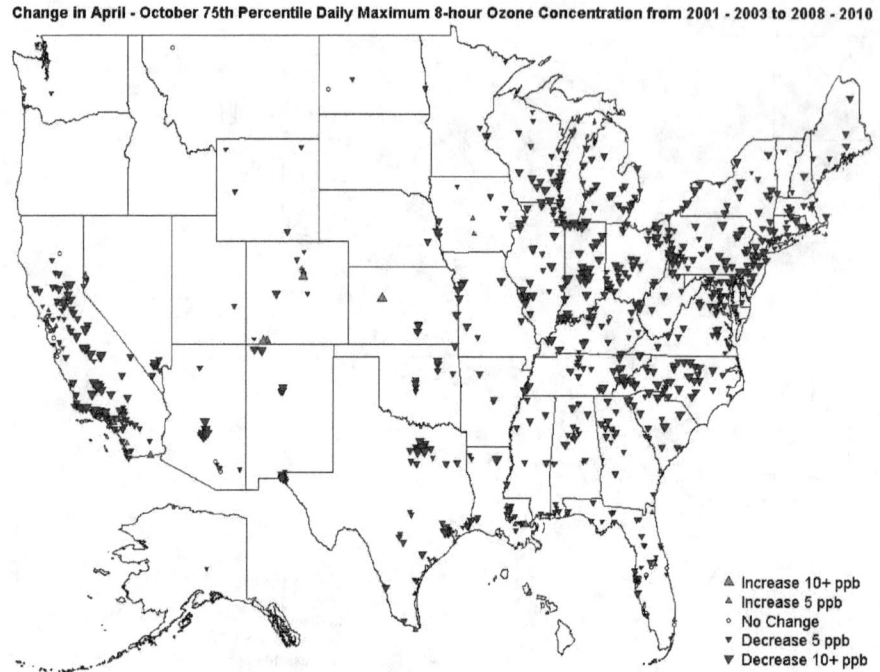

Figure 9: Change in 75th percentile April-October summer season daily 8-hour maximum ozone concentrations between 2001-2003 and 2008-2010.

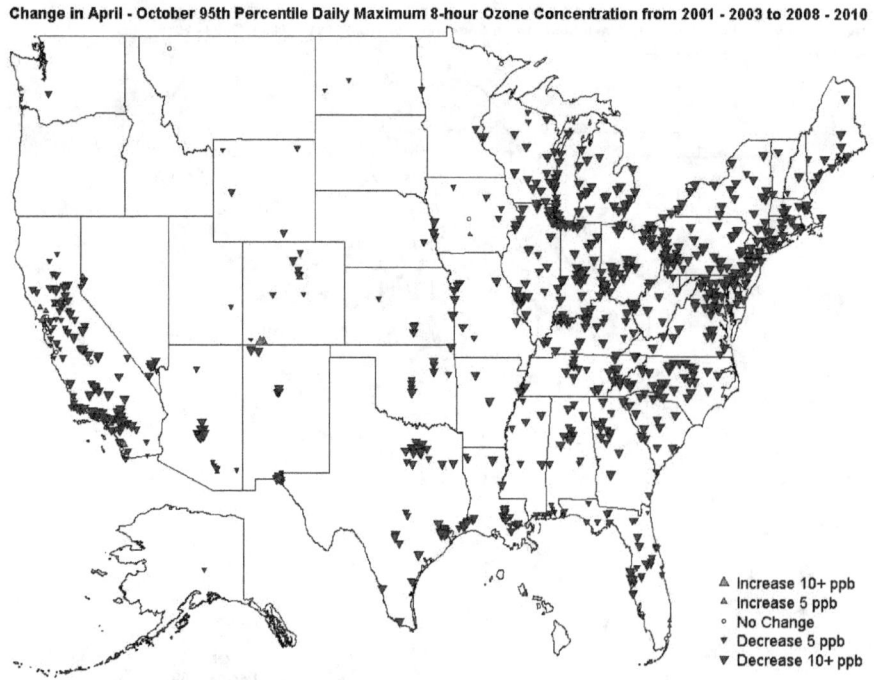

Figure 10: Change in 95th percentile April-October summer season daily 8-hour maximum ozone concentrations between 2001-2003 and 2008-2010.

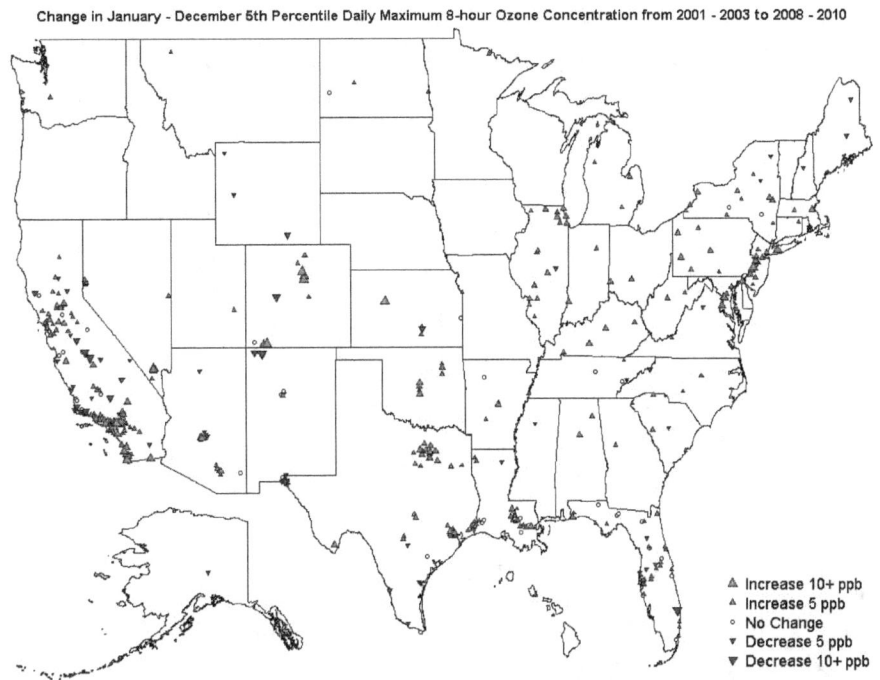

Figure 11: Change in 5th percentile annual daily 8-hour maximum ozone concentrations between 2001-2003 and 2008-2010.

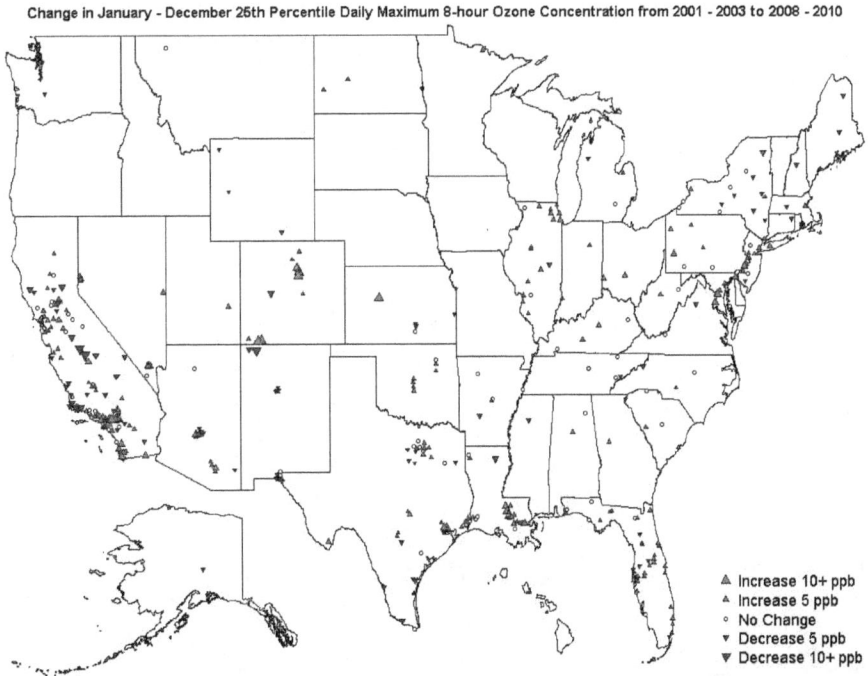

Figure 12: Change in 25th percentile annual daily 8-hour maximum ozone concentrations between 2001-2003 and 2008-2010.

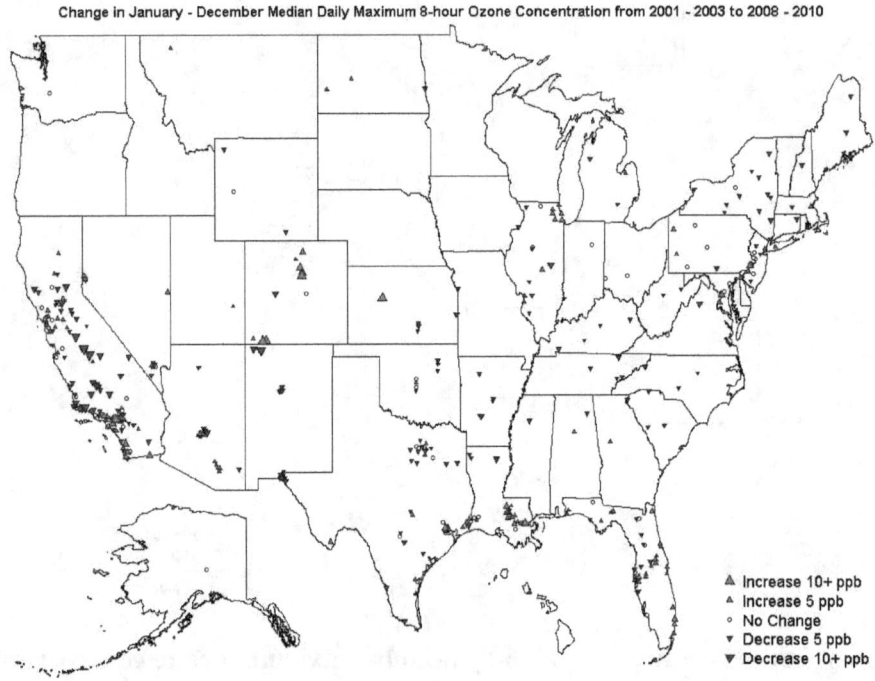

Figure 13: Change in 50th percentile annual daily 8-hour maximum ozone concentrations between 2001-2003 and 2008-2010.

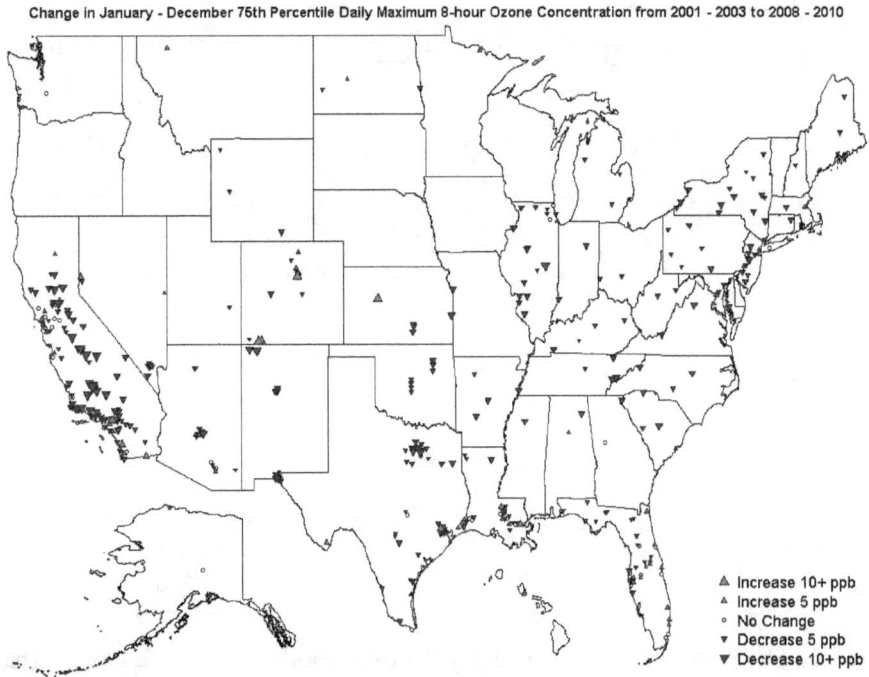

Figure 14: Change in 75th percentile annual daily 8-hour maximum ozone concentrations between 2001-2003 and 2008-2010.

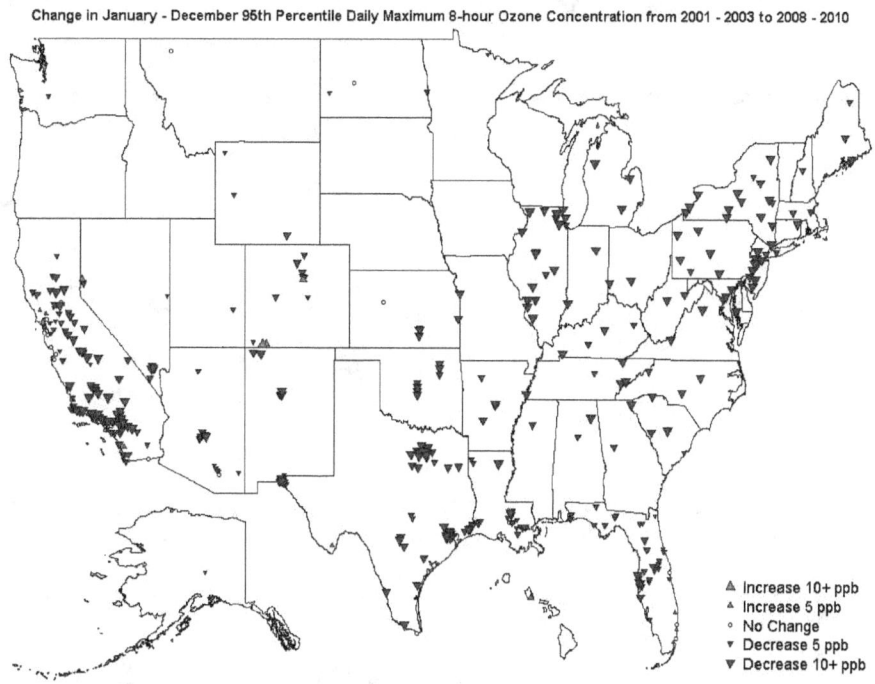

Change in January - December 95th Percentile Daily Maximum 8-hour Ozone Concentration from 2001 - 2003 to 2008 - 2010

△ Increase 10+ ppb
▲ Increase 5 ppb
○ No Change
▽ Decrease 5 ppb
▼ Decrease 10+ ppb

Figure 15: Change in 95th percentile annual daily 8-hour maximum ozone concentrations between 2001-2003 and 2008-2010.

1.2 THIRTEEN-YEAR OZONE TRENDS ACROSS THE COUNTRY AND IN CASE-STUDY AREAS

An initial illustrative summary of the O_3 trends by the categories described in Section 8.2.4 of the main text is shown in Figure 16, where the trend for annual medians of each monitor under study are displayed as separate lines. Although it generally illustrates the range in which average concentrations of O_3 tend to fall (often 40-60 ppb), the simplicity of the plot makes it difficult to discern either spatial or temporal trends. Information about other parts of the annual distribution are also likely to be useful. To concisely display many different distributions in the same template of panels as Figure 16, kernel density estimates (KDEs) of the data were calculated. This process is displayed in Figure 17.

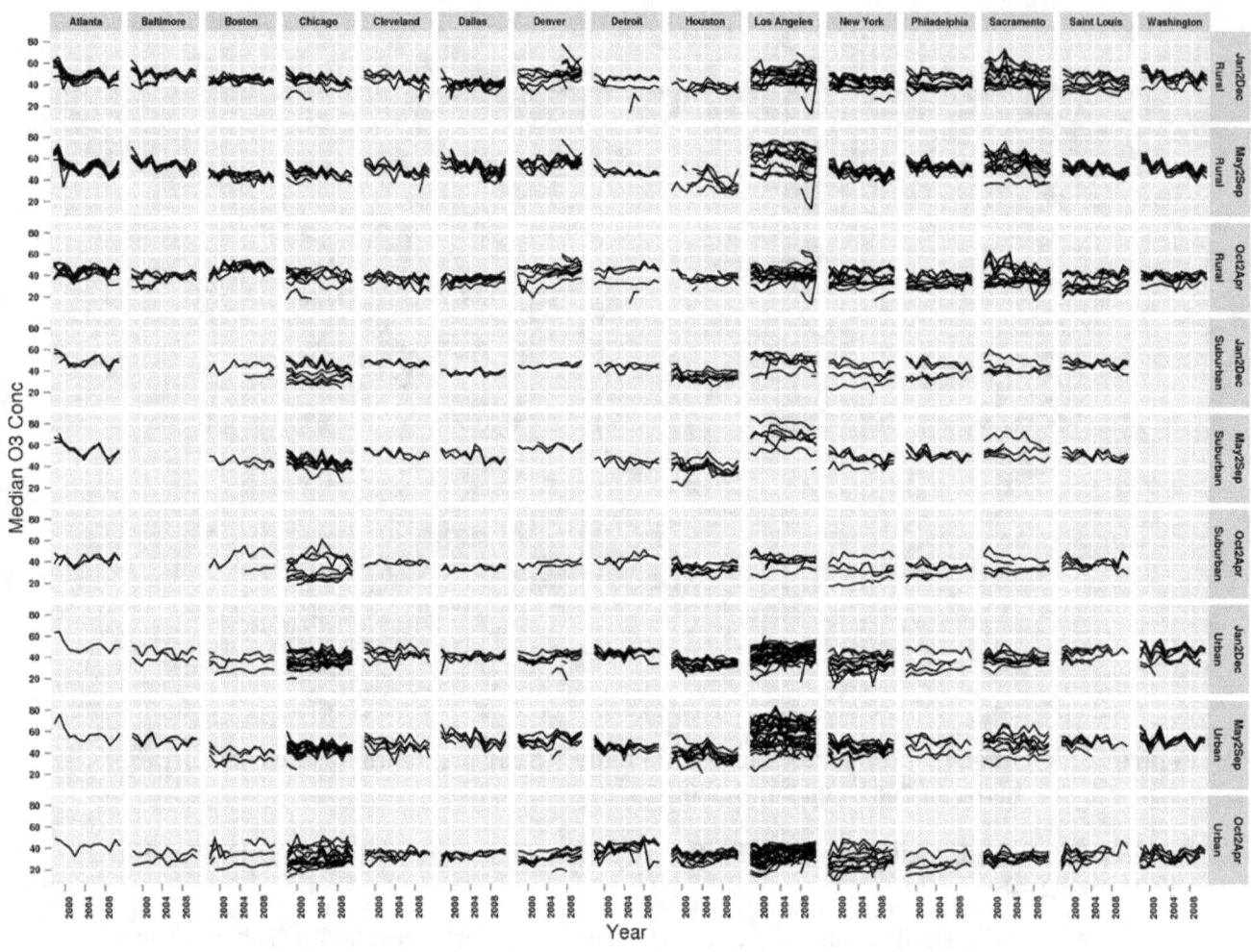

Figure 16: Annual medians of O3 concentrations at each monitor based on different subsets of months.

Figure 17 visually illustrates the process of forming and display a KDE from a year of O₃ data from a single monitoring site. This raw data is displayed in the top panel as a time series of O₃ concentrations. A KDE is then formed from the raw data, which is similar in principle to a histogram, which gives counts of data that fall within user-defined bins. However, the KDE "smoothes" out the histogram so that arbitrary bins do not need to be set, and converts the counts to a "density". The density can yield a probability if desired, but that is beyond the scope of this display; for our purposes, a higher density for a given concentration simply means that more O₃ measurements were collected near that value compared to other possible concentrations. The curve of the KDE can then be converted to a color stripe as shown in the bottom panel of Figure 17, where the color is related to the height of the curve in the middle panel.

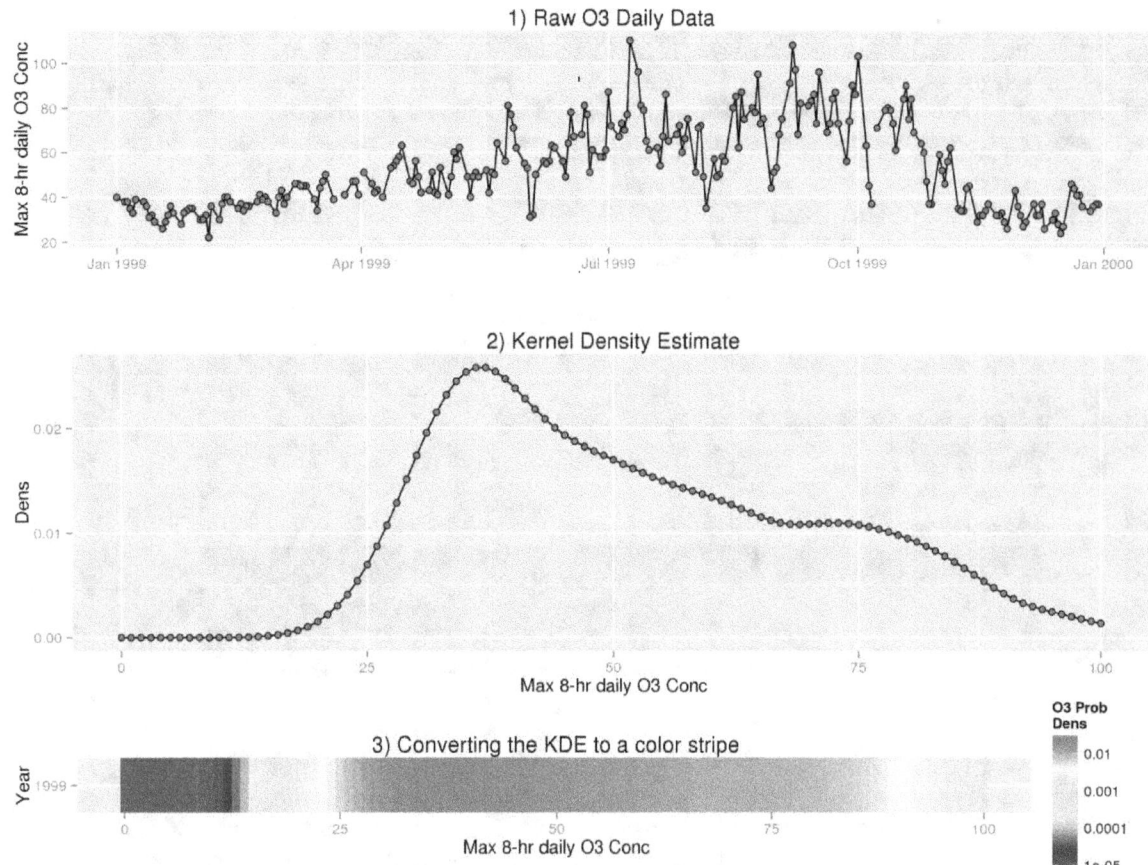

Figure 17: Procedure for creating the display of O3 distributions shown in Figure 18

Each year of data shown in the groups in Figure 16 was thus converted to a color-based KDE as shown in Figure 17, and the resulting collection of KDEs is shown in Figure 18. Annual medians and modes of the distributions across all monitors in each group indicated by the plot's panels are also shown, with color indicating the direction of the trend over time. Statistical significance for multi-year ozone trends was determined using the Spearman rank order correlation coefficient (p-value < 0.05). The general pattern of KDEs over time appears to be either small or insignificant changes to the central tendencies of the distributions (i.e. mode and median), but with a "condensing" of the concentration to the 40-50 ppb range, meaning that lower concentrations grow and high concentrations decrease.

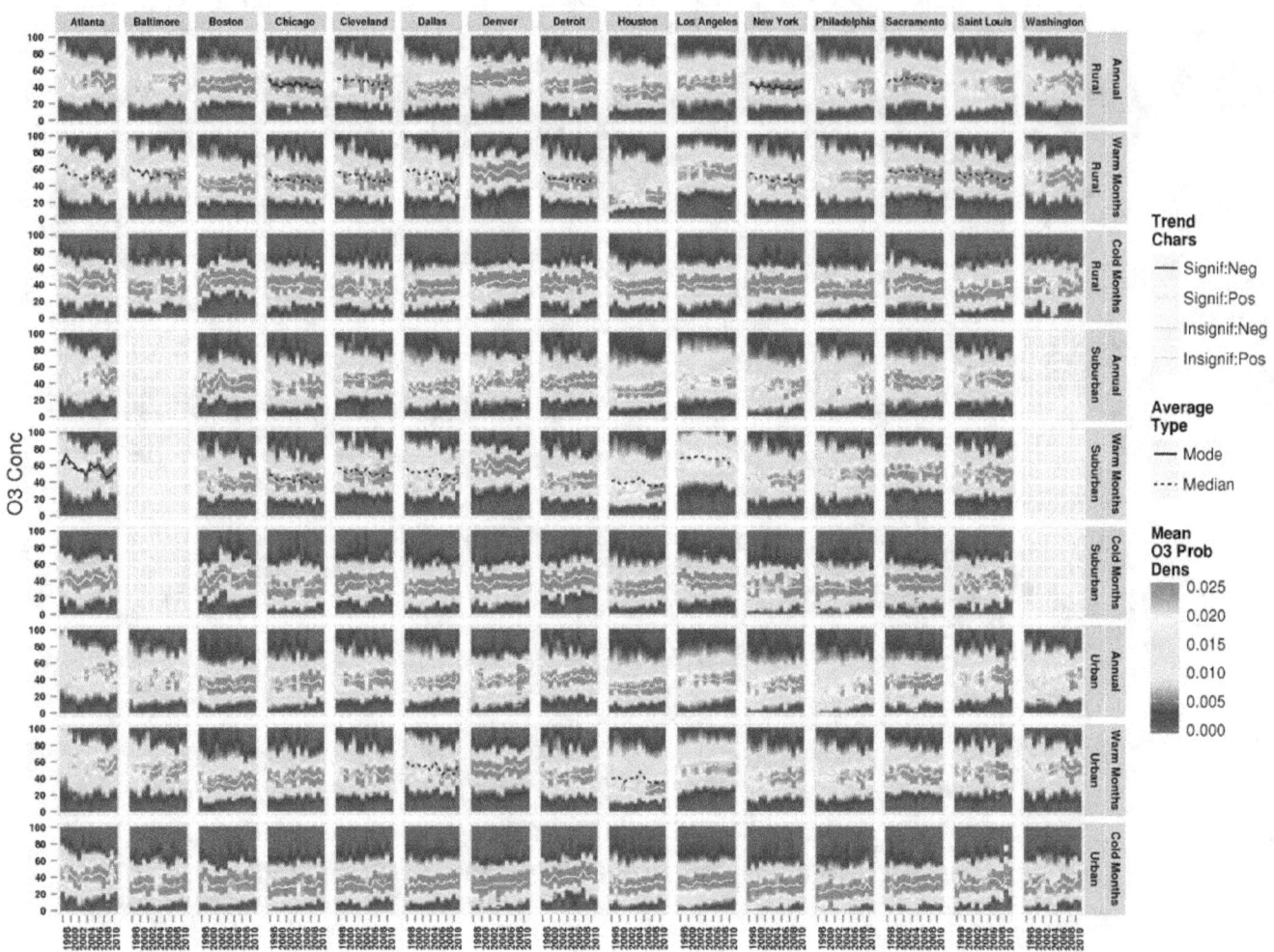

Figure 18: KDEs of groups of monitors' annual O3 concentrations for different subsets of months, shown on a linear color scale. The modes and medians of these concentrations across the year and monitors for each group are shown in the overlaying lines.

Section 8.2.3 in the main text provided maps showing summertime (May-September) ozone trends at specific monitor locations within two urban case study areas. Here, we provide similar maps for the other 13 case study areas. In section 8.2.3 we described the general trend of fourth high ozone values decreasing in most locations while mean and median values were more likely to increase in core urban areas and decrease in surrounding suburban and rural areas. In addition, in most cities, the monitor with the highest design values did not occur in the urban core. These trends were demonstrated by maps of the New York and Chicago areas in the main text. Here we see that the trends are visible in many other urban case study areas, including Baltimore, Boston, Cleveland, Denver, Houston, Los Angeles, and Saint Louis. However, ozone trends in a few urban areas exhibit different behavior. In Atlanta and Sacramento, the highest design value monitor occurs near the urban core. In Atlanta, mid-range ozone has statistically significant decreases trends at monitors both in the urbanized and in the surrounding area. All urban monitors in Detroit and Sacramento showed no significant trend in either mean or median ozone values. In Dallas, significant increases in mid-range ozone occurred at sites outside of the urban core. Finally in Philadelphia, there was no statistically significant trend at any monitor for the fourth highest 8-hour daily maximum ozone value.

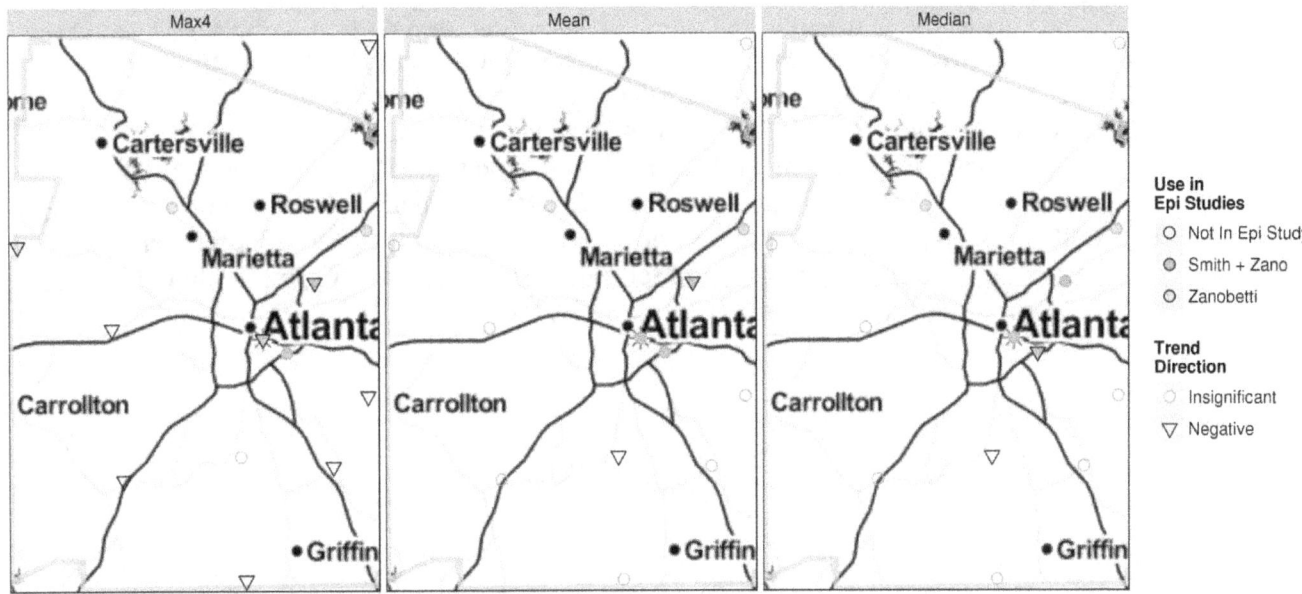

Figure 19: Map of ozone trends at specific monitor locations in the Atlanta area. All upward and downward facing triangles represent statistically significant trends from 1998-2001 (p<0.05), circles represent locations with no significant trends. The pink star indicates the site with the highest design values in 2011. Left panel shows trends in May-September 4th highest 8-hour daily maximum ozone values, center panel shows trends in May-September mean 8-hour daily maximum, and left panel shows trends in May-September median 8-hour daily maximum ozone values.

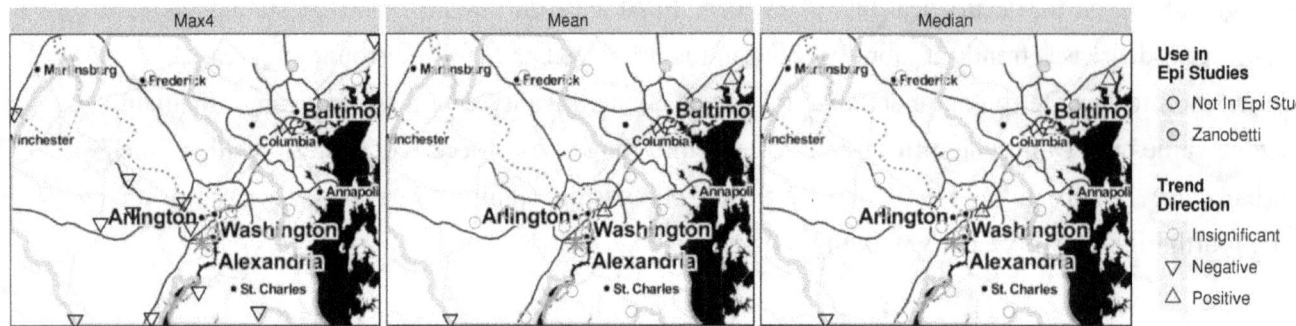

Figure 20: Map of ozone trends at specific monitor locations in the Baltimore/Washington D.C. area. All upward and downward facing triangles represent statistically significant trends from 1998-2001 (p<0.05), circles represent locations with no significant trends. The pink star indicates the site with the highest design values in 2011. Left panel shows trends in May-September 4th highest 8-hour daily maximum ozone values, center panel shows trends in May-September mean 8-hour daily maximum, and left panel shows trends in May-September median 8-hour daily maximum ozone values.

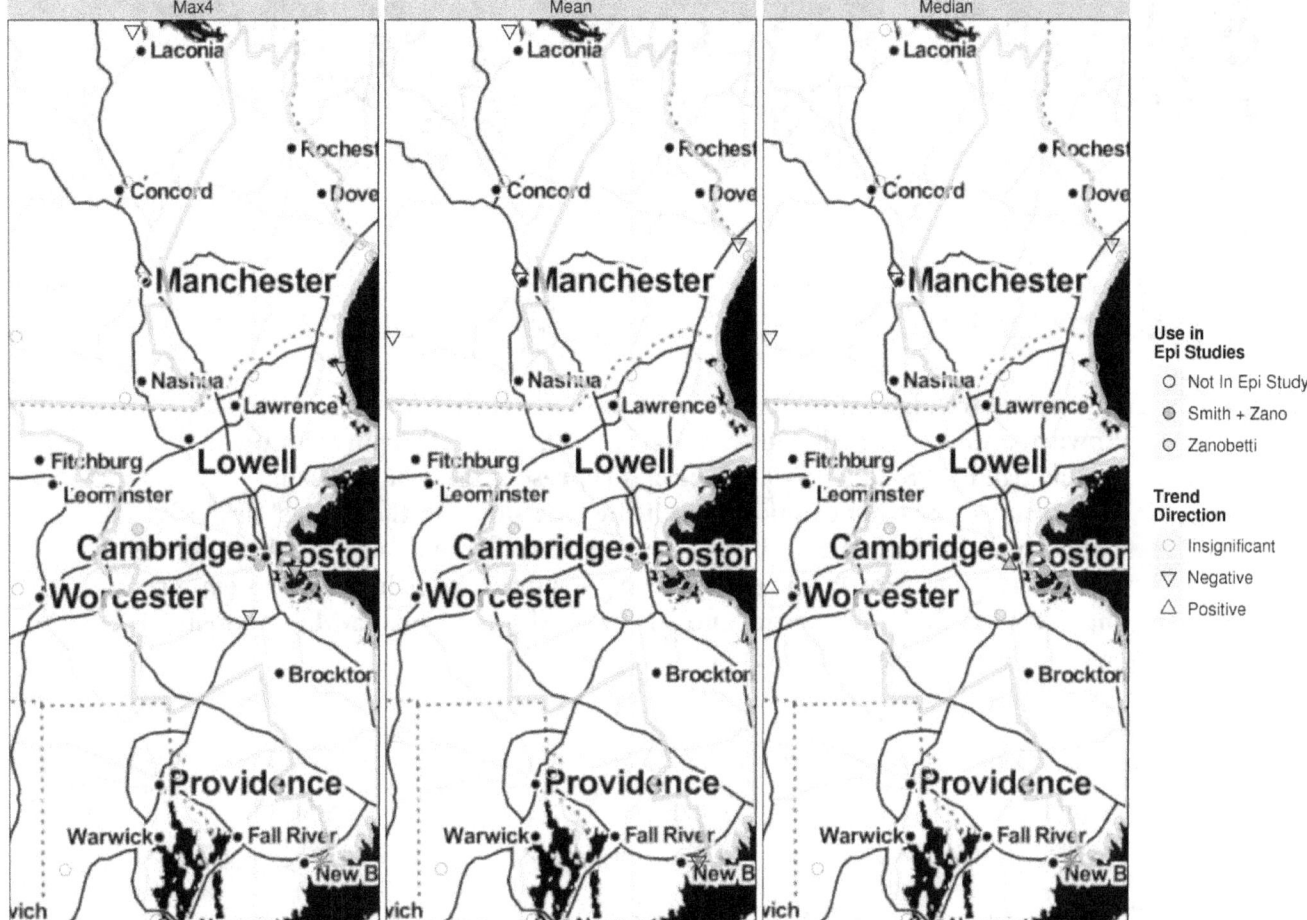

Figure 21: Map of ozone trends at specific monitor locations in the Boston area. All upward and downward facing triangles represent statistically significant trends from 1998-2001 (p<0.05), circles represent locations with no significant trends. The pink star indicates the site with the highest design values in 2011. Left panel shows trends in May-September 4th highest 8-hour daily maximum ozone values, center panel shows trends in May-September mean 8-hour daily maximum, and left panel shows trends in May-September median 8-hour daily maximum ozone values.

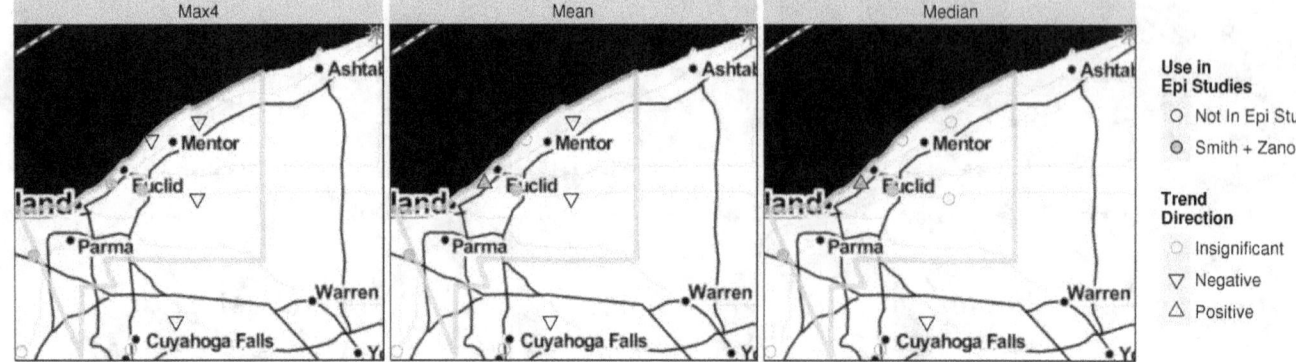

Figure 22: Map of ozone trends at specific monitor locations in the Cleveland area. All upward and downward facing triangles represent statistically significant trends from 1998-2001 (p<0.05), circles represent locations with no significant trends. The pink star indicates the site with the highest design values in 2011. Left panel shows trends in May-September 4th highest 8-hour daily maximum ozone values, center panel shows trends in May-September mean 8-hour daily maximum, and left panel shows trends in May-September median 8-hour daily maximum ozone values.

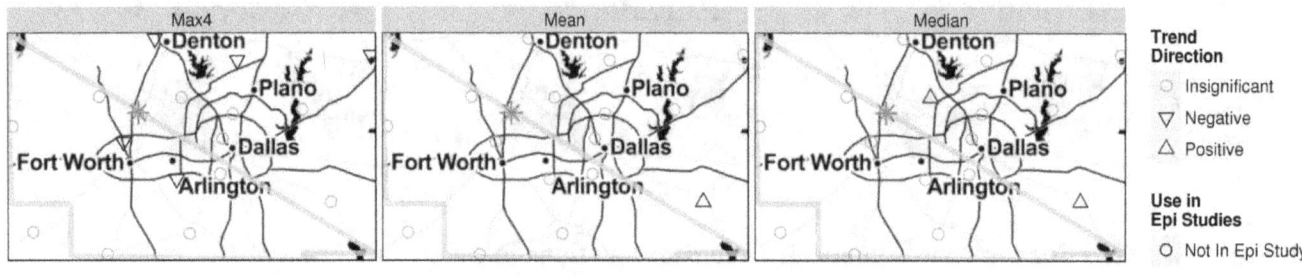

Figure 23: Map of ozone trends at specific monitor locations in the Dallas area. All upward and downward facing triangles represent statistically significant trends from 1998-2001 (p<0.05), circles represent locations with no significant trends. The pink star indicates the site with the highest design values in 2011. Left panel shows trends in May-September 4th highest 8-hour daily maximum ozone values, center panel shows trends in May-September mean 8-hour daily maximum, and left panel shows trends in May-September median 8-hour daily maximum ozone values.

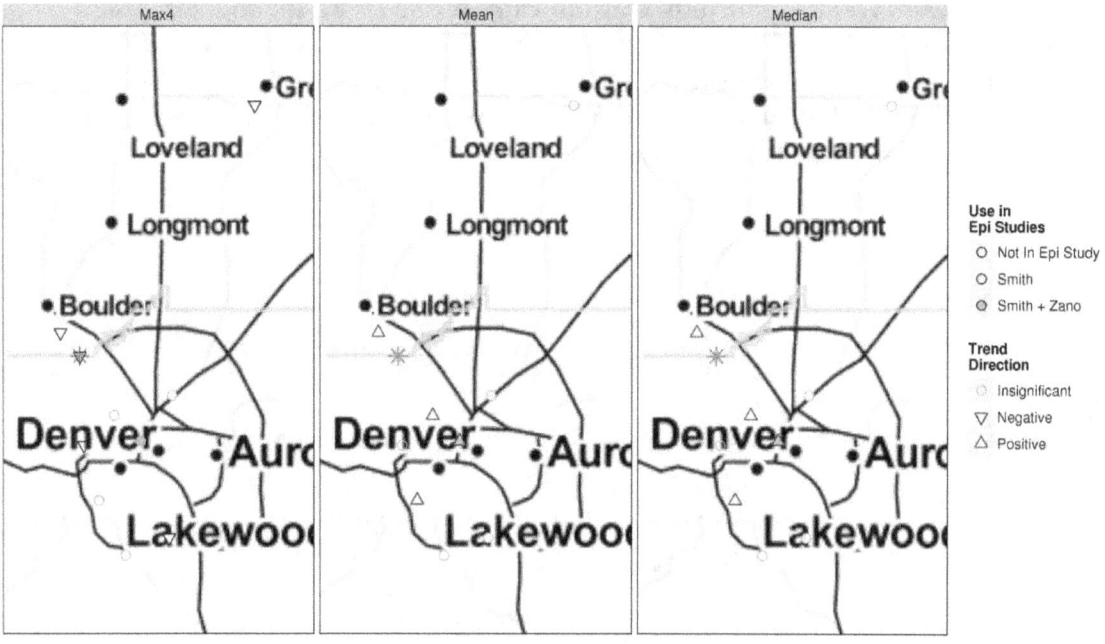

Figure 24: Map of ozone trends at specific monitor locations in the Denver area. All upward and downward facing triangles represent statistically significant trends from 1998-2001 (p<0.05), circles represent locations with no significant trends. The pink star indicates the site with the highest design values in 2011. Left panel shows trends in May-September 4th highest 8-hour daily maximum ozone values, center panel shows trends in May-September mean 8-hour daily maximum, and left panel shows trends in May-September median 8-hour daily maximum ozone values.

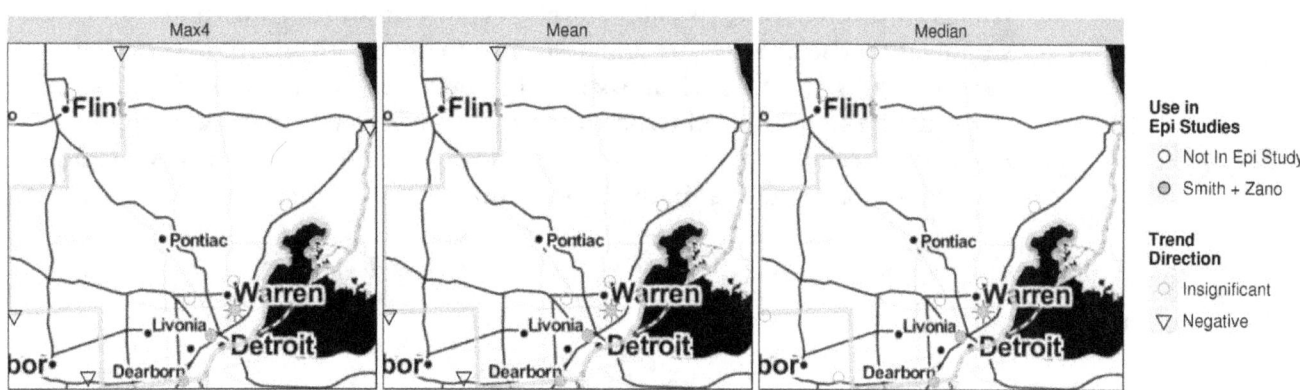

Figure 25: Map of ozone trends at specific monitor locations in the Detroit area. All upward and downward facing triangles represent statistically significant trends from 1998-2001 (p<0.05), circles represent locations with no significant trends. The pink star indicates the site with the highest design values in 2011. Left panel shows trends in May-September 4th highest 8-hour daily maximum ozone values, center panel shows trends in May-September mean 8-hour daily maximum, and left panel shows trends in May-September median 8-hour daily maximum ozone values.

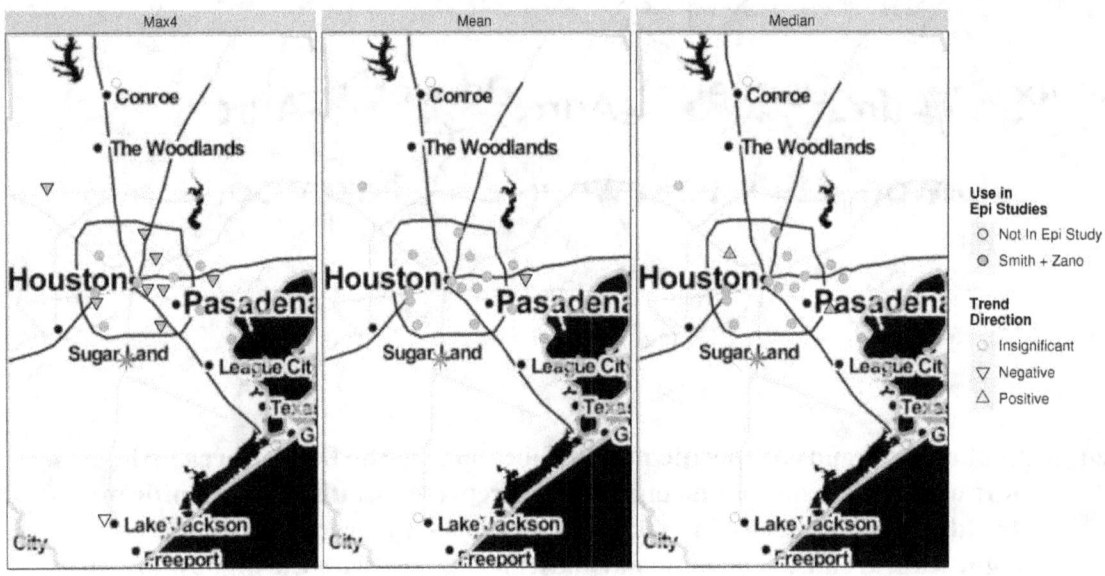

Figure 26: Map of ozone trends at specific monitor locations in the Houston area. All upward and downward facing triangles represent statistically significant trends from 1998-2001 (p<0.05), circles represent locations with no significant trends. The pink star indicates the site with the highest design values in 2011. Left panel shows trends in May-September 4th highest 8-hour daily maximum ozone values, center panel shows trends in May-September mean 8-hour daily maximum, and left panel shows trends in May-September median 8-hour daily maximum ozone values.

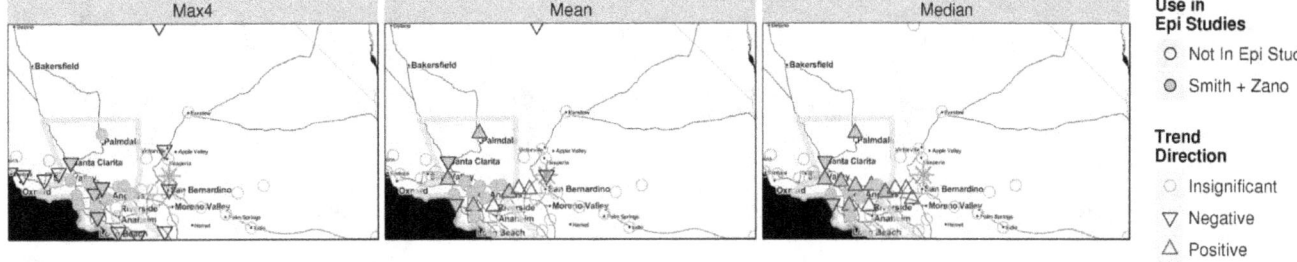

Figure 27: Map of ozone trends at specific monitor locations in the Los Angeles area. All upward and downward facing triangles represent statistically significant trends from 1998-2001 (p<0.05), circles represent locations with no significant trends. The pink star indicates the site with the highest design values in 2011. Left panel shows trends in May-September 4th highest 8-hour daily maximum ozone values, center panel shows trends in May-September mean 8-hour daily maximum, and left panel shows trends in May-September median 8-hour daily maximum ozone values.

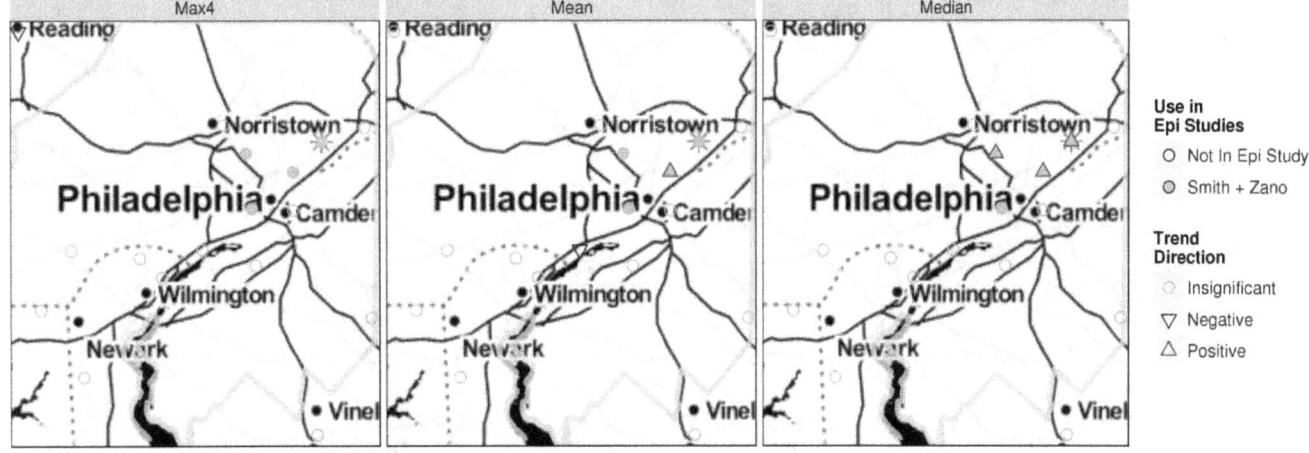

Figure 28: Map of ozone trends at specific monitor locations in the Philadelphia area. All upward and downward facing triangles represent statistically significant trends from 1998-2001 (p<0.05), circles represent locations with no significant trends. The pink star indicates the site with the highest design values in 2011. Left panel shows trends in May-September 4th highest 8-hour daily maximum ozone values, center panel shows trends in May-September mean 8-hour daily maximum, and left panel shows trends in May-September median 8-hour daily maximum ozone values.

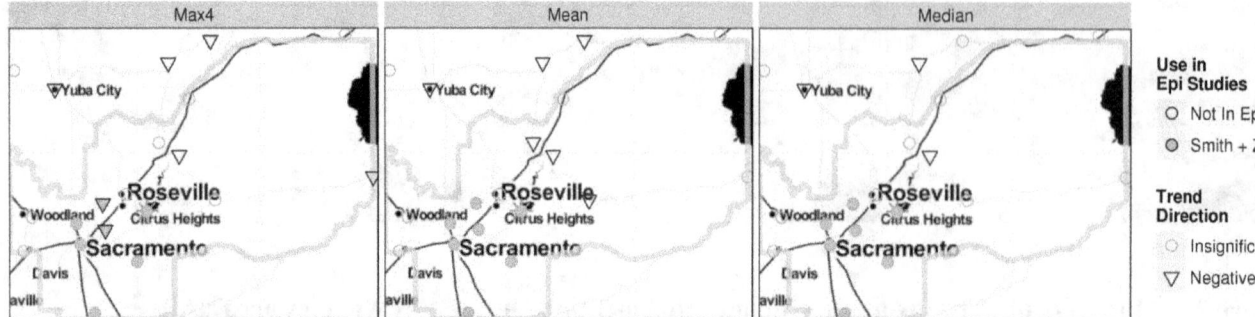

Figure 29: Map of ozone trends at specific monitor locations in the Sacramento area. All upward and downward facing triangles represent statistically significant trends from 1998-2001 (p<0.05), circles represent locations with no significant trends. The pink star indicates the site with the highest design values in 2011. Left panel shows trends in May-September 4th highest 8-hour daily maximum ozone values, center panel shows trends in May-September mean 8-hour daily maximum, and left panel shows trends in May-September median 8-hour daily maximum ozone values.

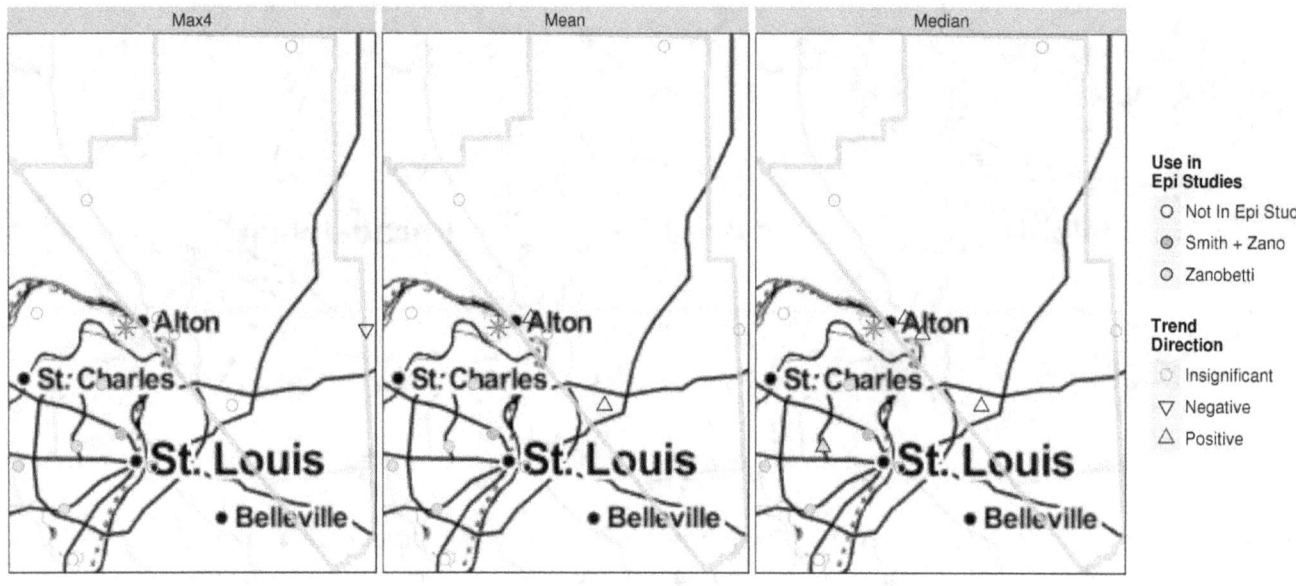

Figure 30: Map of ozone trends at specific monitor locations in the Saint Louis area. All upward and downward facing triangles represent statistically significant trends from 1998-2001 (p<0.05), circles represent locations with no significant trends. The pink star indicates the site with the highest design values in 2011. Left panel shows trends in May-September 4th highest 8-hour daily maximum ozone values, center panel shows trends in May-September mean 8-hour daily maximum, and left panel shows trends in May-September median 8-hour daily maximum ozone values.

In addition to the ozone trends, Chapter 8 includes Table 8.7 which shows relationships between regional trends in NOx and VOC emissions and regional ozone trends. The objective was to investigate possible similarities in broad trends of O_3 concentrations and anthropogenic NOx and VOC emissions. Trends of emissions were derived from county-level emissions data from the 2002, 2005, 2008, and 2011 EPA National Emissions Inventory (NEI). This data was in the form of annual totals for the 'Tier 1' sectors, which refers to the most general classification scheme of source categories in the NEI. This raw data is plotted in U.S. maps in Figure 31. The row of maps labeled "TierTotal" refers to the sum of all the other maps. Note that the Wildfires and Biogenics sectors are absent from all these analyses due to their large magnitude and non-anthropogenic origin.

To analyze trends, emissions were spatially summed for each year and each sector across the NOAA Climate Regions[1] (shown in Figure 32). The resulting trend lines for each sector and emissions pollutant are shown in Figure 33. For direct comparison to O3 trends, the ozone data from the study areas was grouped together by the same NOAA climate regions, and annual percentiles of the resulting distributions were calculated, which are shown in Figure 34 and Figure 35. The descriptors show in Table 8.7 of the main document were derived from Figure 33, Figure 34, and Figure 35.

---

[1] Climate regions are defined by NOAA's National Climate Data Center: http://www ncdc noaa.gov/monitoring-references/maps/us-climate-regions.php

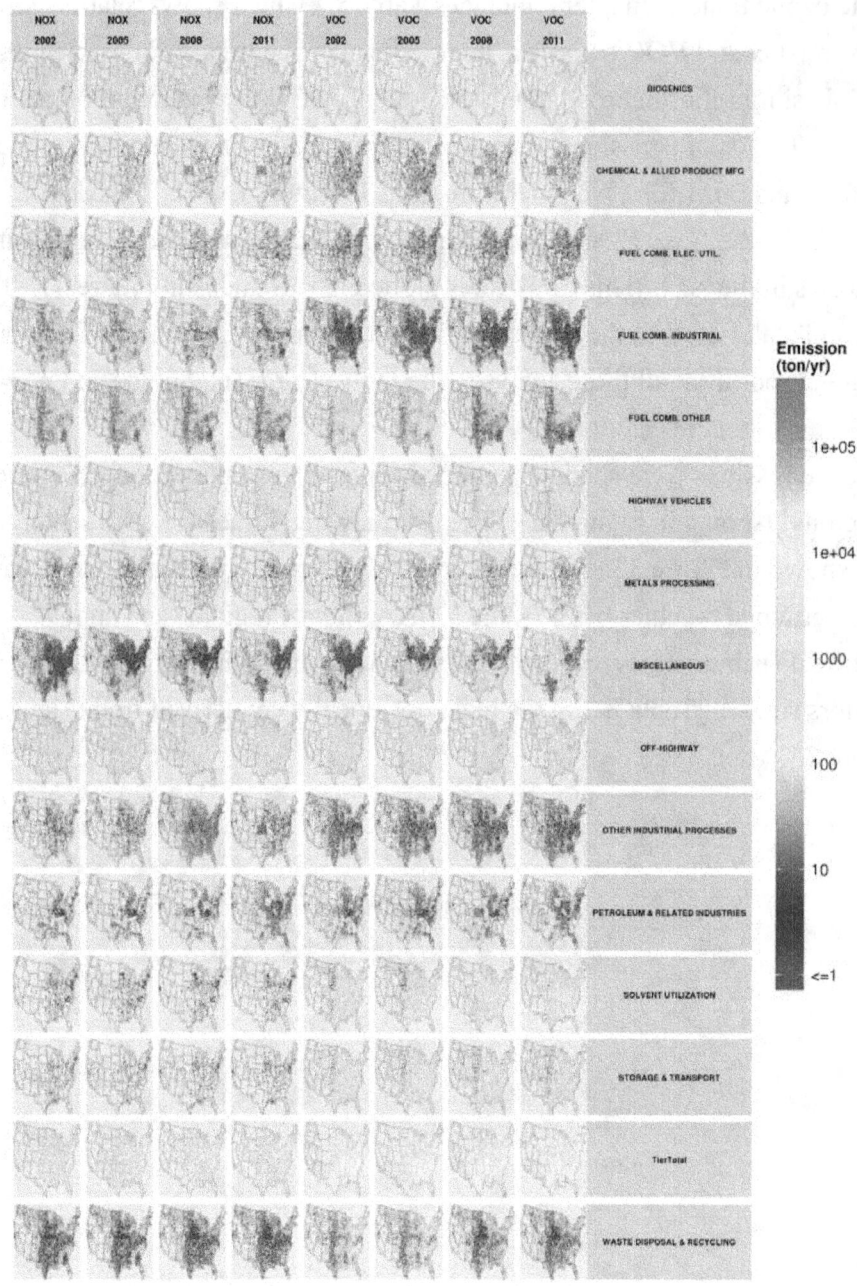

Figure 31: Maps of NOx and VOC emissions by source sector for 2002, 2005, 2008, and 2011

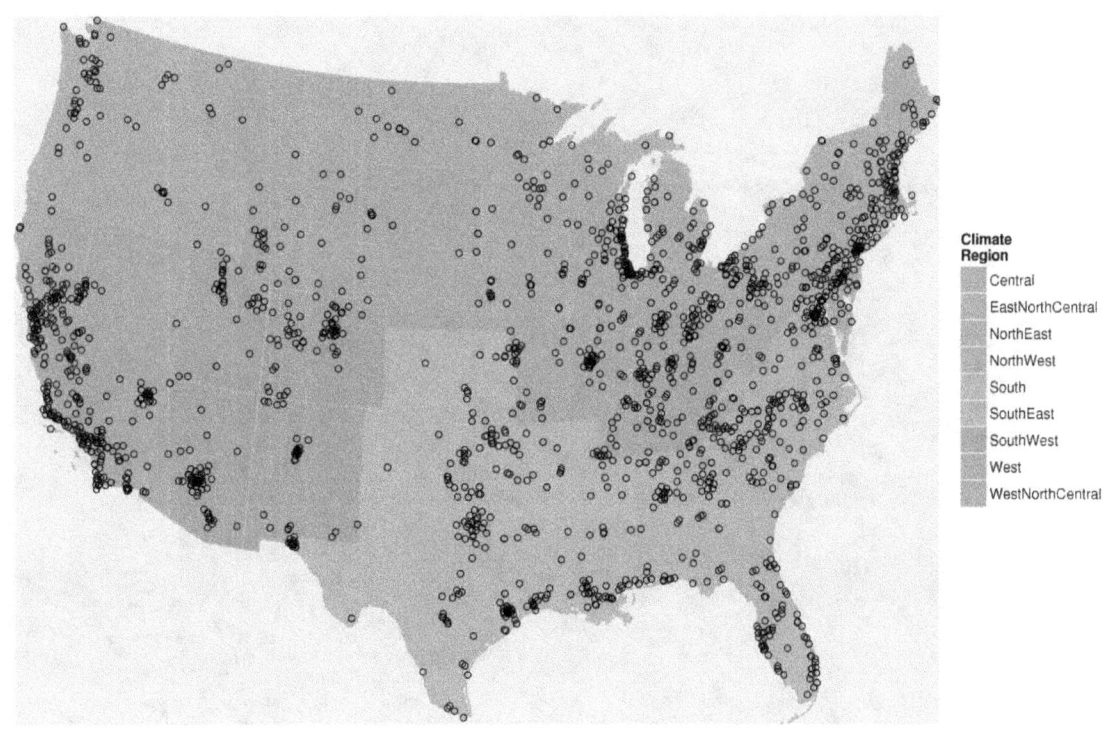

Figure 32: Map of nine NOAA climate regions that were used to aggregate emissions and ambient ozone trends. Dots show locations of ozone monitors.

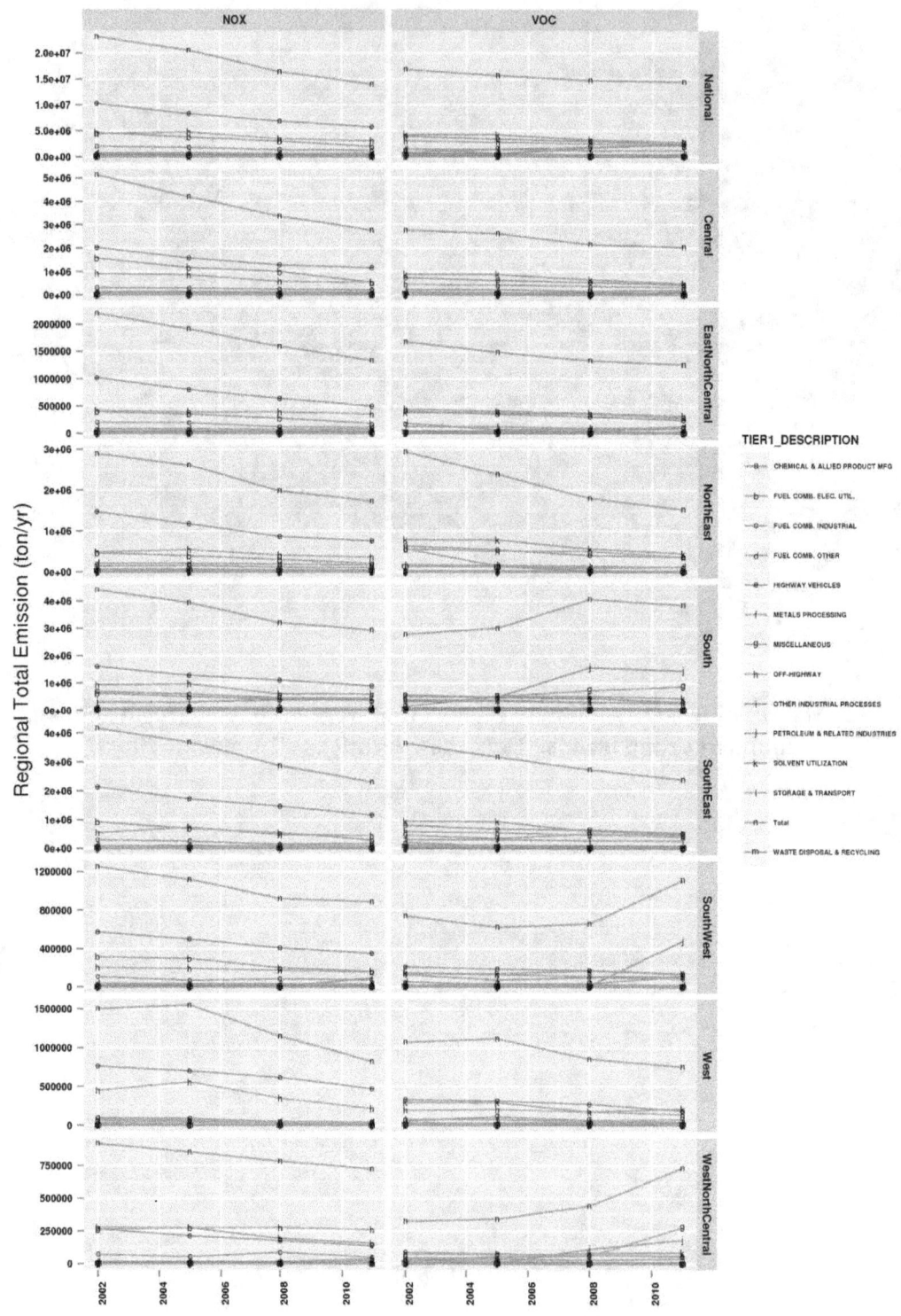

Figure 33: Plots of NOx and VOC emissions trends by source sector. Emissions are aggregated by NOAA climate region and by urban, rural, and suburban location.

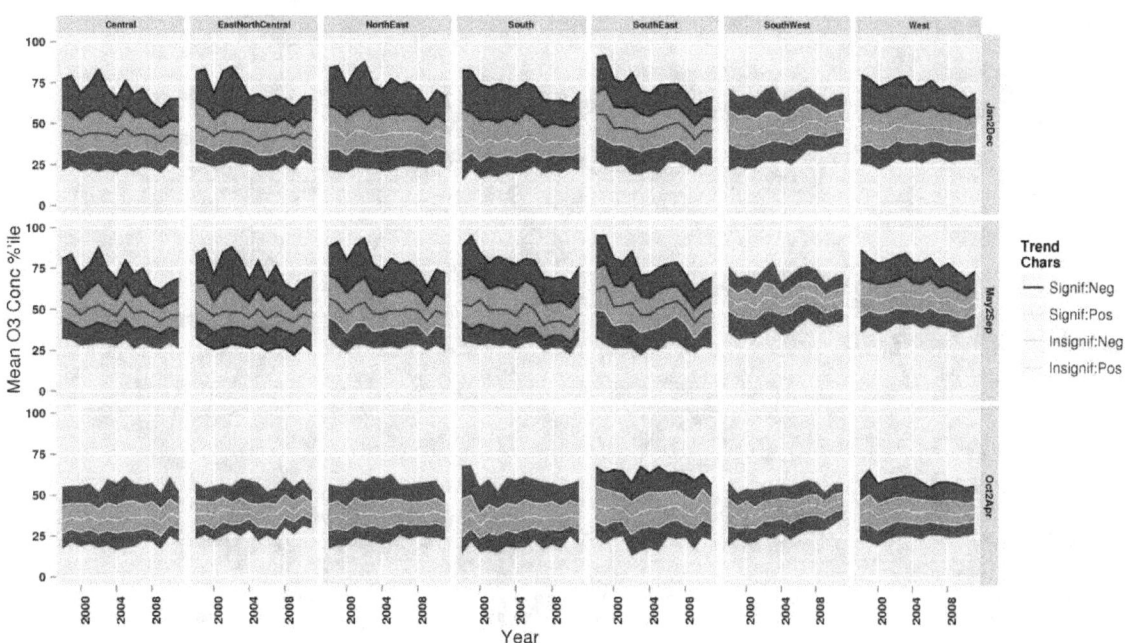

Figure 34: Distributions of low population density (rural) monitors' O₃ concentrations for different subsets of months over a 13-year period. From top to bottom in each ribbon plot, the blue and white lines indicate the spatial mean of the 95th, 75th, 50th, 25th, and 5th percentiles for each monitor for every year from 1998-2011. Trend are shown for 7 of 9 NOAA climate regions (The Northwest and West North Central regions did not contain any case-study areas).

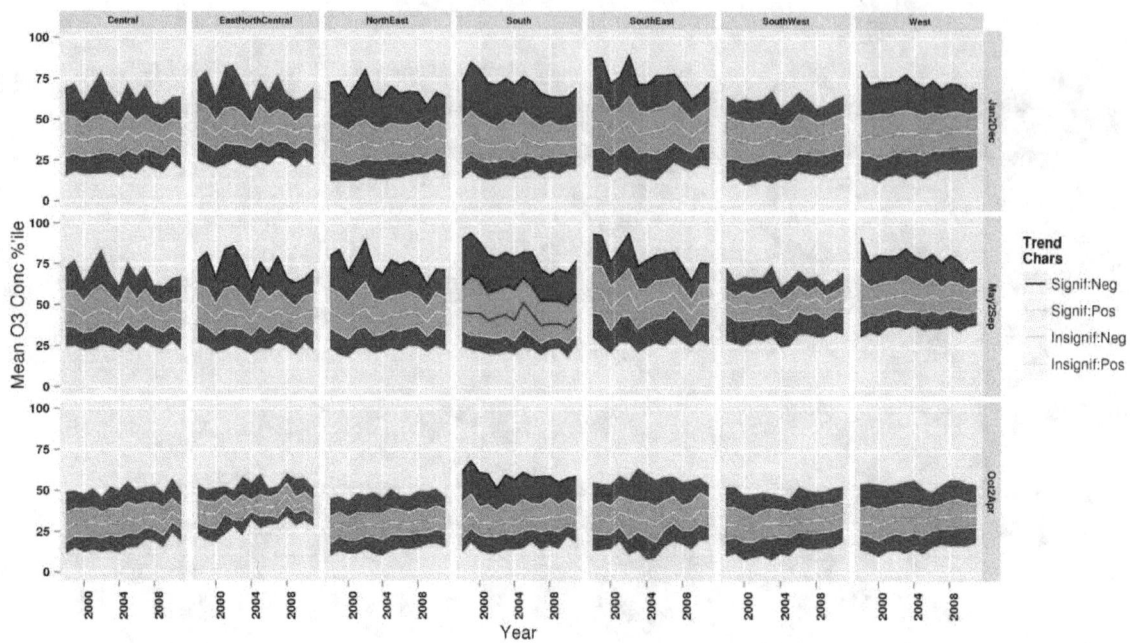

Figure 35: Distributions of high population density monitors' O₃ concentrations for different subsets of months over a 13-year period. From top to bottom in each ribbon plot, the blue and white lines indicate the spatial mean of the 95th, 75th, 50th, 25th, and 5th percentiles for each monitor for every year from 1998-2011. Trend are shown for each of 7 of 9 NOAA climate regions (The Northwest and West North Central regions did not contain any case-study areas).

2. MODELED OZONE CHANGES IN RESPONSE TO ACROSS THE BOARD EMISSIONS REDUCTIONS

2.1 MAPS OF RATIOS OF MEAN OZONE FROM 2007 CMAQ SIMULATIONS INCLUDING EMISSIONS REDUCTIONS TO MEAN OZONE FROM 2007 BASE CMAQ SIMULATIONS.

In section 8.2.3.2 we evaluated ozone response from two CMAQ model simulations with across-the-board reductions in US anthropogenic emissions. We presented results using ratios of the mean ozone concentrations in the emissions reduction scenario to mean ozone concentrations in the 2007 base CMAQ simulation. Here we provide a full set of maps which include mean ozone response over three different time periods (January 2007, April-October 2007, and May-September 2007) and for four different emissions reduction scenarios (50% NOx reductions, 50% NOx and VOC reductions, 90% NOx reductions, and 90% NOx and VOC reductions). These plots show that ozone increases are most pronounced in cooler months with January maps showing broad ozone increases across most of the modeling domain while May-September maps show broad ozone decreases across most of the modeling domain. The April-October maps show ozone decreases in most areas but localized increases in some large cities. Also, comparing the NOx and VOC reductions to reductions in NOx alone show the VOC has some effect at decreasing region ozone but does not fully mitigate ozone increases in urban areas in the April-October maps nor change the general trends described above.

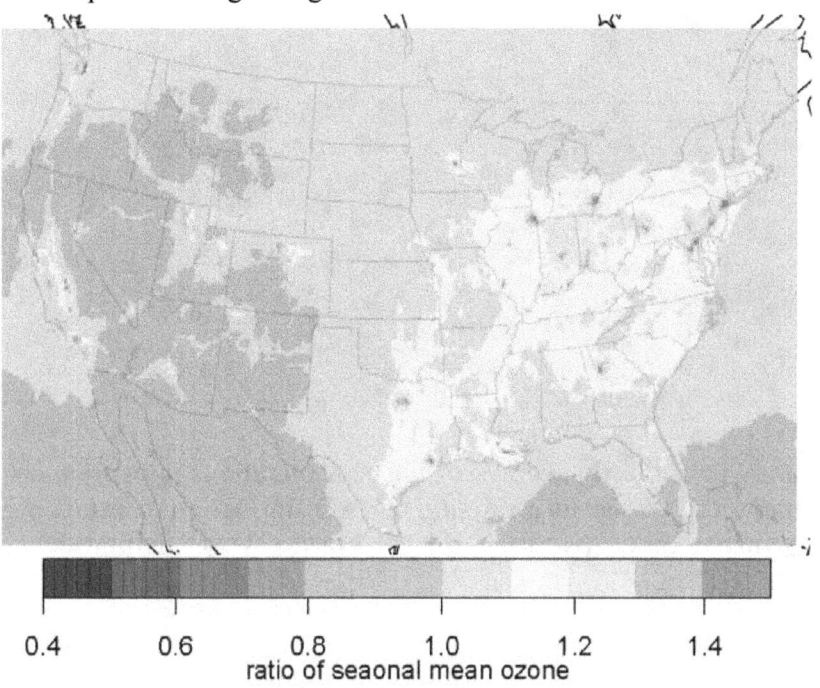

Figure 36: Ratio of January monthly average ozone concentrations in brute force 50% NOx emissions reduction CMAQ simulation to January monthly average ozone concentration in the 2007 base CMAQ simulation.

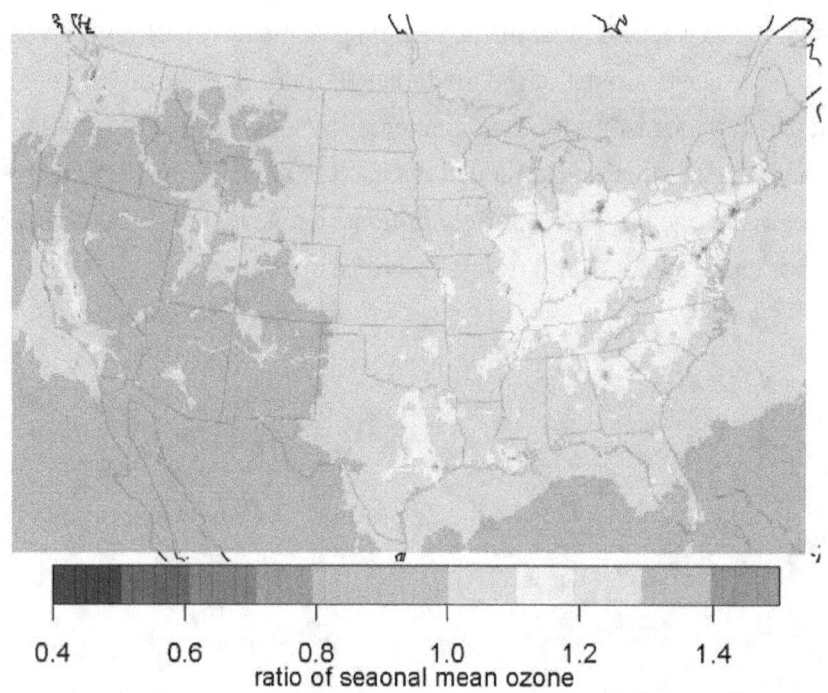

Figure 37: Ratio of January monthly average ozone concentrations in brute force 50% NOx and VOC emissions reduction CMAQ simulation to January monthly average ozone concentration in the 2007 base CMAQ simulation.

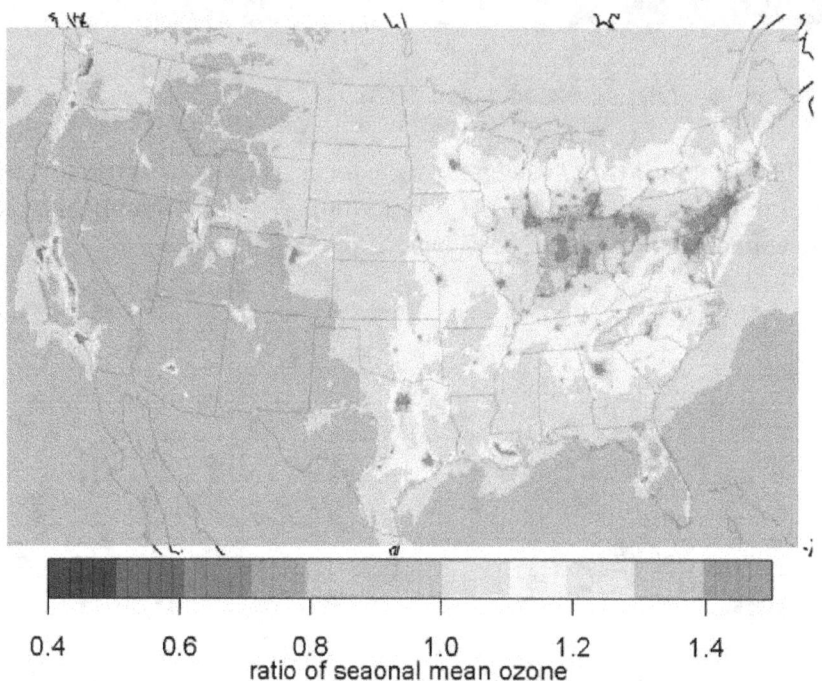

Figure 38: Ratio of January monthly average ozone concentrations in brute force 50% NOx and VOC emissions reduction CMAQ simulation to January monthly average ozone concentration in the 2007 base CMAQ simulation.

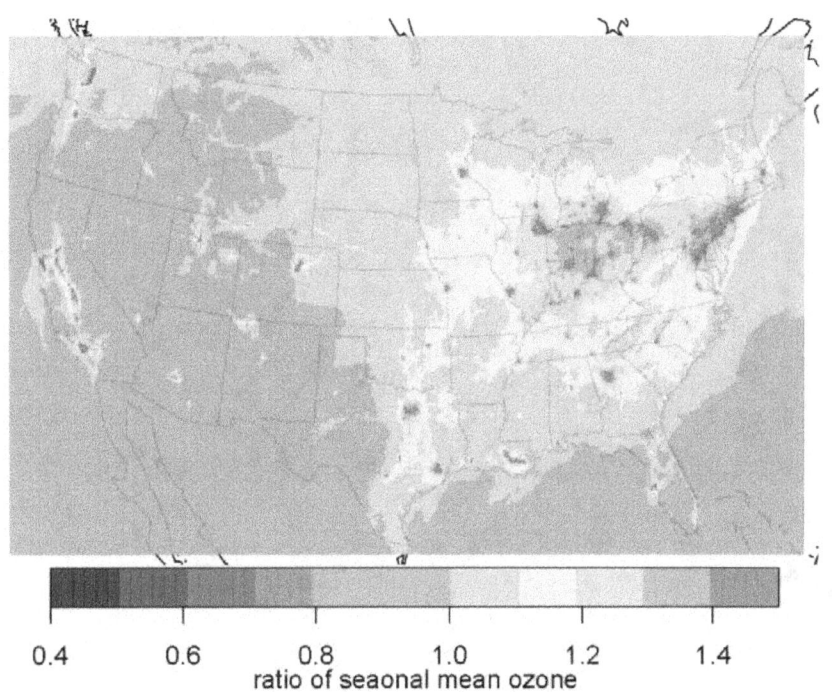

0.4   0.6   0.8   1.0   1.2   1.4
ratio of seaonal mean ozone

Figure 39: Ratio of January monthly average ozone concentrations in brute force 90% NOx and emissions reduction CMAQ simulation to January monthly average ozone concentration in the 2007 base CMAQ simulation.

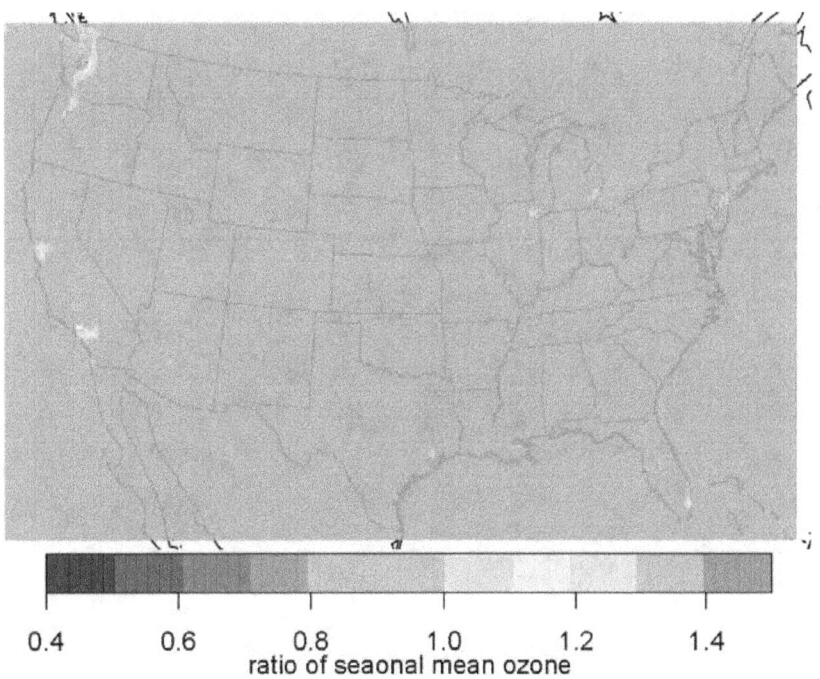

0.4   0.6   0.8   1.0   1.2   1.4
ratio of seaonal mean ozone

Figure 40: Ratio of April-October seasonal average ozone concentrations in brute force 50% NOx emissions reduction CMAQ simulation to April-October seasonal average ozone concentration in the 2007 base CMAQ simulation.

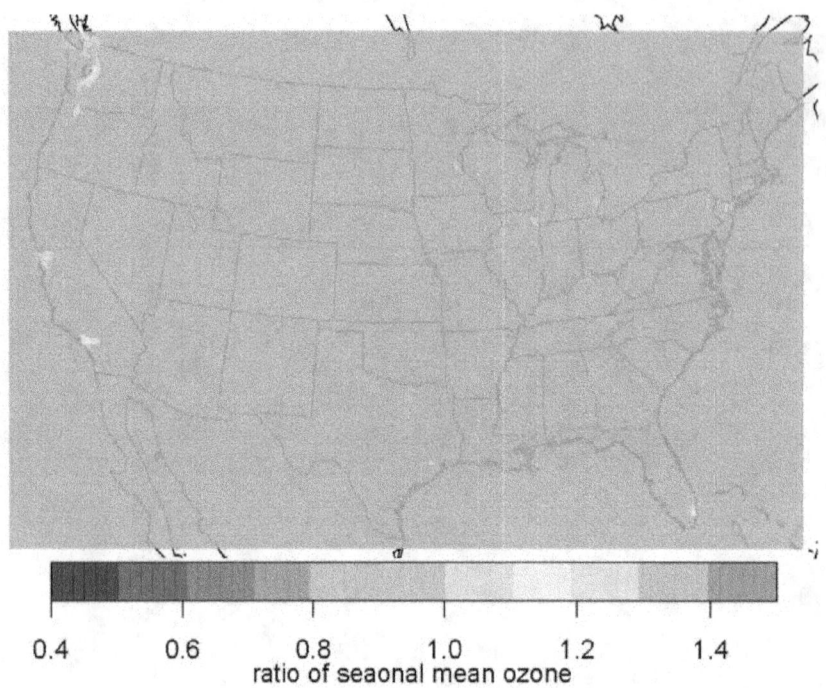

0.4　　0.6　　0.8　　1.0　　1.2　　1.4
ratio of seaonal mean ozone

Figure 41: Ratio of April-October seasonal average ozone concentrations in brute force 50% NOx and VOC emissions reduction CMAQ simulation to April-October seasonal average ozone concentration in the 2007 base CMAQ simulation.

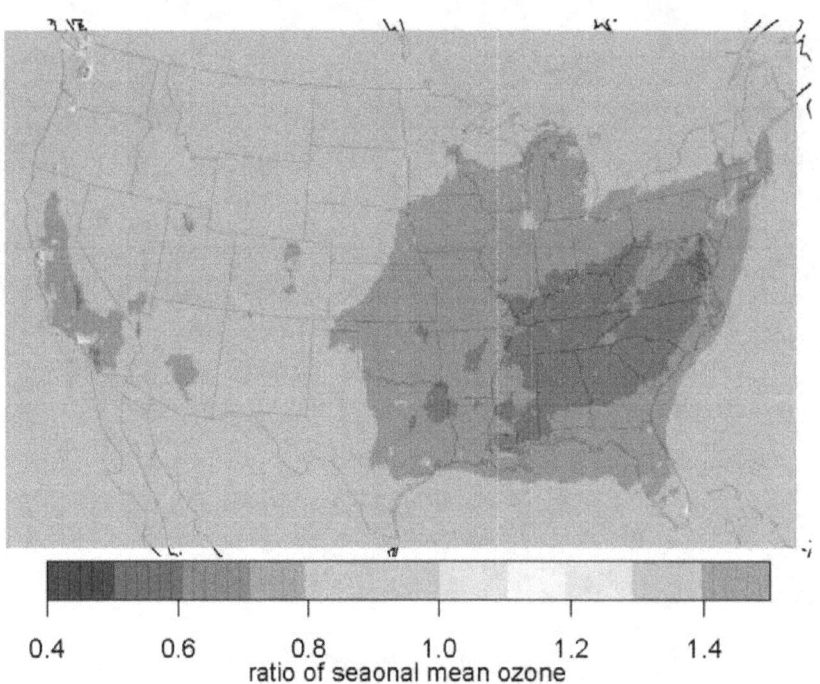

0.4　　0.6　　0.8　　1.0　　1.2　　1.4
ratio of seaonal mean ozone

Figure 42: Ratio of April-October seasonal average ozone concentrations in brute force 90% NOx emissions reduction CMAQ simulation to April-October seasonal average ozone concentration in the 2007 base CMAQ simulation.

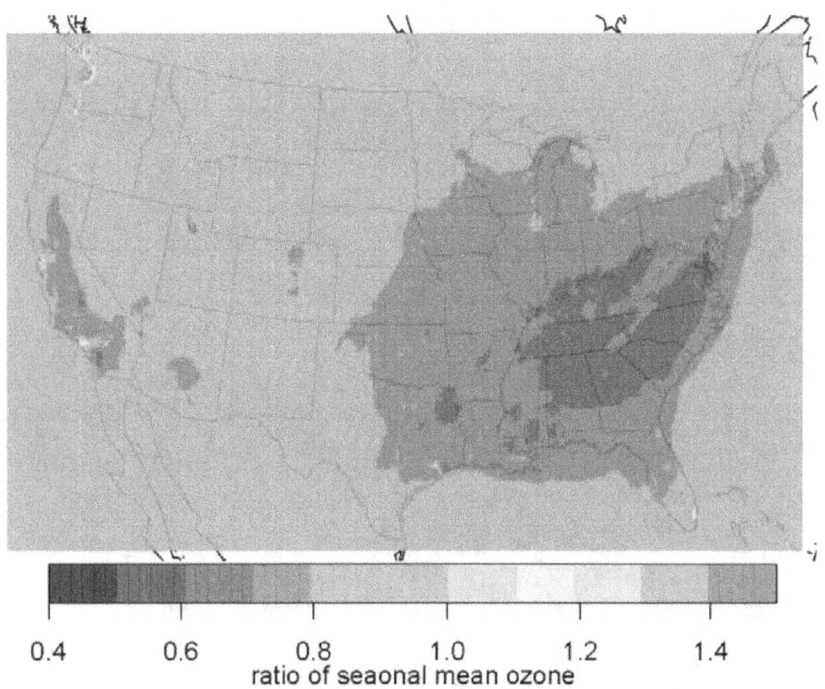

Figure 43: Ratio of April-October seasonal average ozone concentrations in brute force 90% NOx and VOC emissions reduction CMAQ simulation to April-October seasonal average ozone concentration in the 2007 base CMAQ simulation.

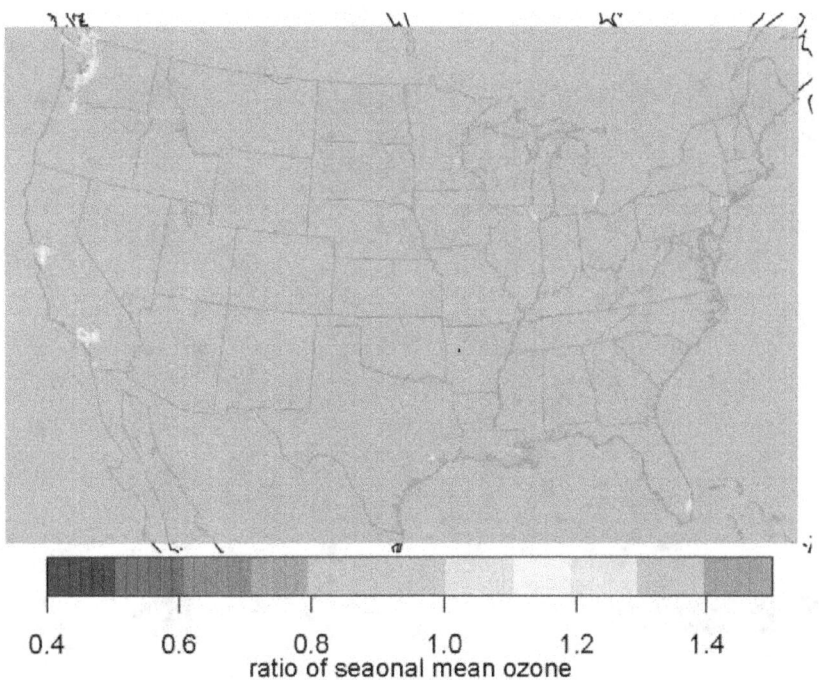

Figure 44: Ratio of May-September seasonal average ozone concentrations in brute force 50% NOx emissions reduction CMAQ simulation to May-September seasonal average ozone concentration in the 2007 base CMAQ simulation.

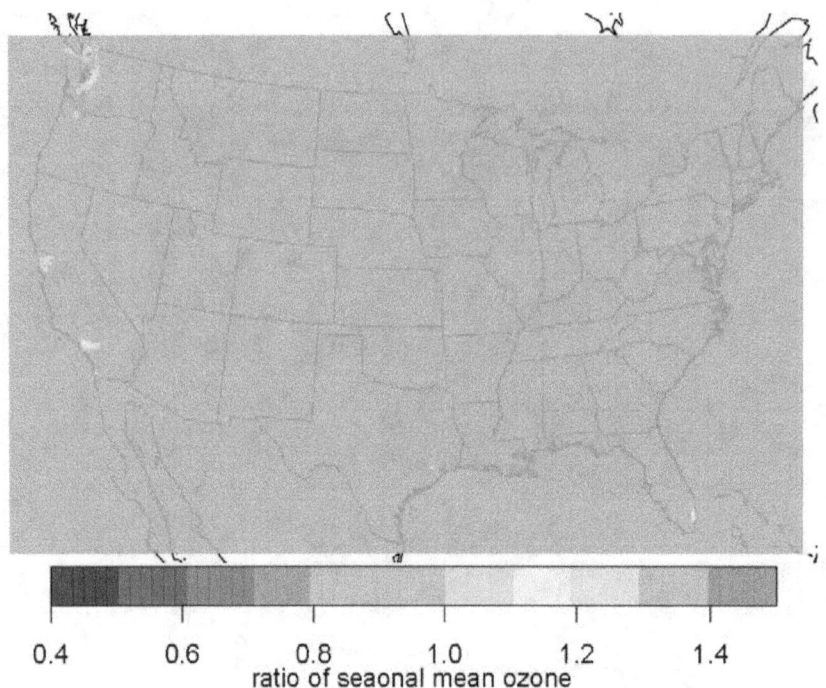

ratio of seaonal mean ozone

Figure 45: Ratio of May-September seasonal average ozone concentrations in brute force 50% NOx and VOC emissions reduction CMAQ simulation to May-September seasonal average ozone concentration in the 2007 base CMAQ simulation.

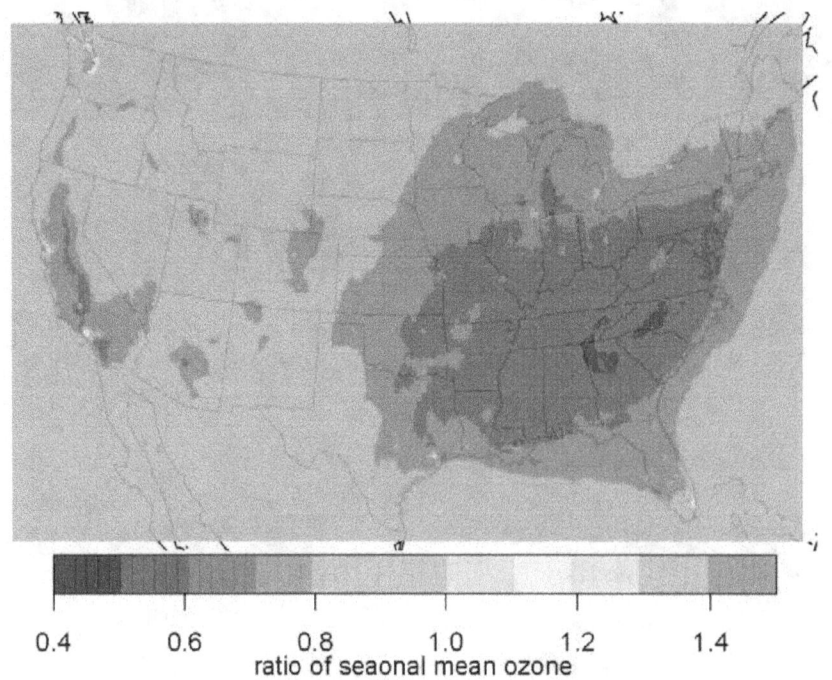

ratio of seaonal mean ozone

Figure 46: Ratio of May-September seasonal average ozone concentrations in brute force 90% NOx emissions reduction CMAQ simulation to May-September seasonal average ozone concentration in the 2007 base CMAQ simulation.

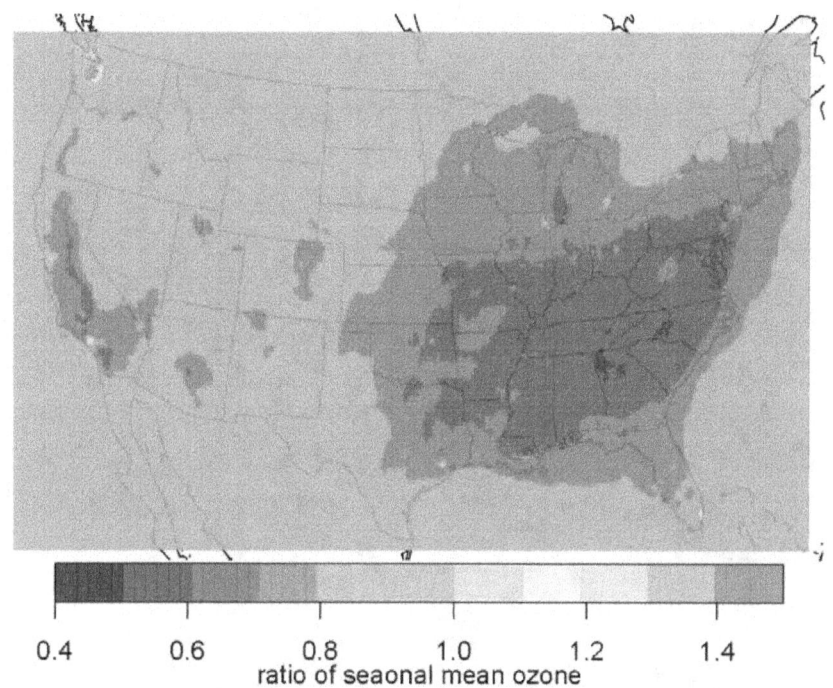

Figure 47: Ratio of May-September seasonal average ozone concentrations in brute force 90% NOx and VOC emissions reduction CMAQ simulation to May-September seasonal average ozone concentration in the 2007 base CMAQ simulation.

These maps can be further understood by breaking down response by month and binning increases and decreases by base ozone concentration. Figure 48 and Figure 49 show this breakdown for each emissions reduction scenario. This plot clearly shows that ozone increases predominantly occur at lower base ozone concentrations while high modeled base ozone concentrations appear to decrease in almost all cases in the emissions reduction scenarios. The ozone decreases occur more often and are more substantial during the months of June, July, August, and September. The 90% NOx reduction simulations have few locations with ozone increases than the 50% NOx reduction simulation however the are a limited number of grid cells in which the ozone increases are larger in the 90% NOx reduction than in the 50% NOx reduction simulation. Overall, the distributions of ozone response look similar when VOC reductions are added on top of NOx reductions, although the NOx and VOC reduction cases are shifted slightly more toward reducing ozone than the NOx only reduction cases.

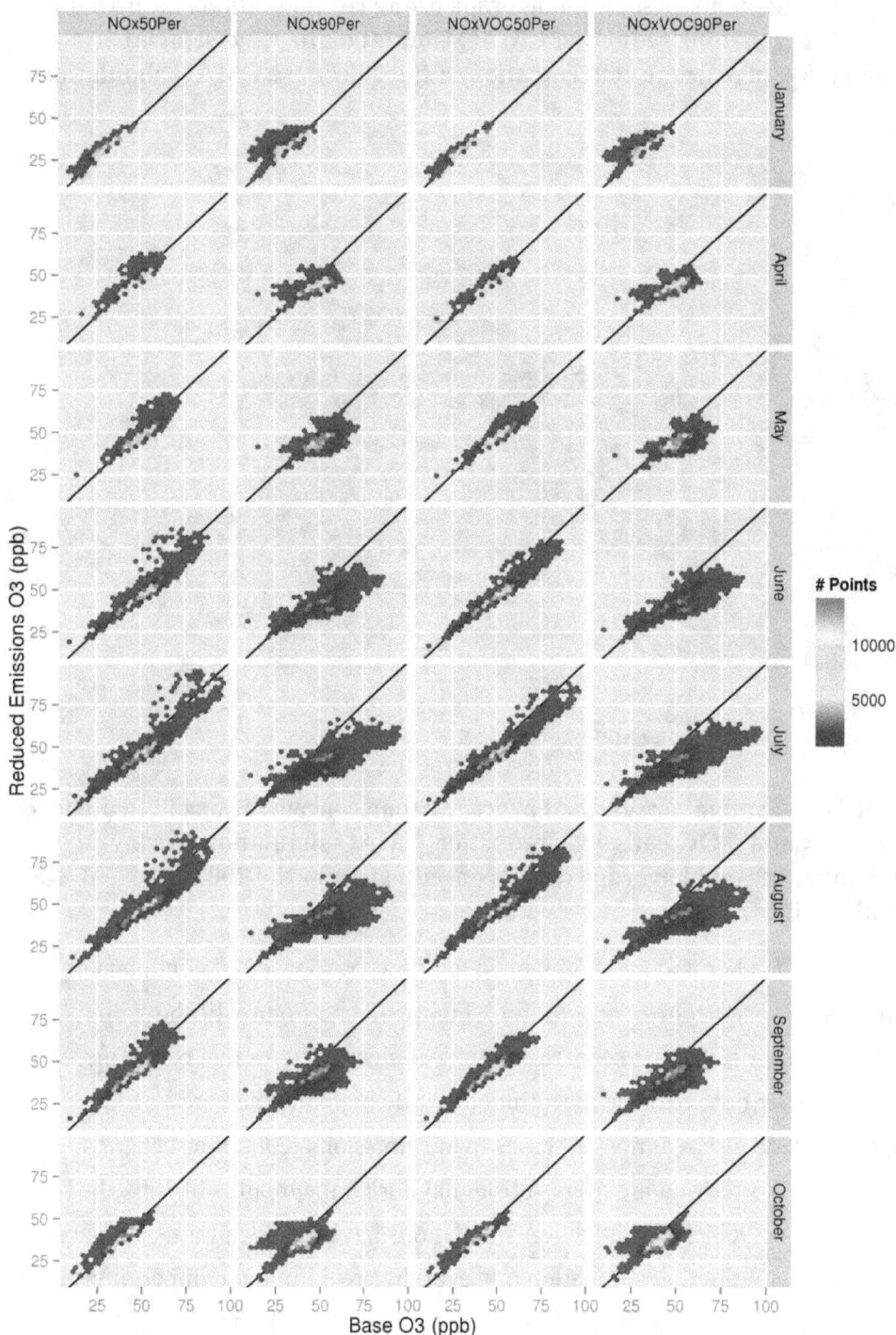

Figure 48: Density scatter plot comparing modeled monthly mean ozone in the 2007 base CMAQ simulation to modeled monthly mean ozone in the emissions reduction CMAQ simulations. Colors depict the number of points occurring at any location on the scatter plot.

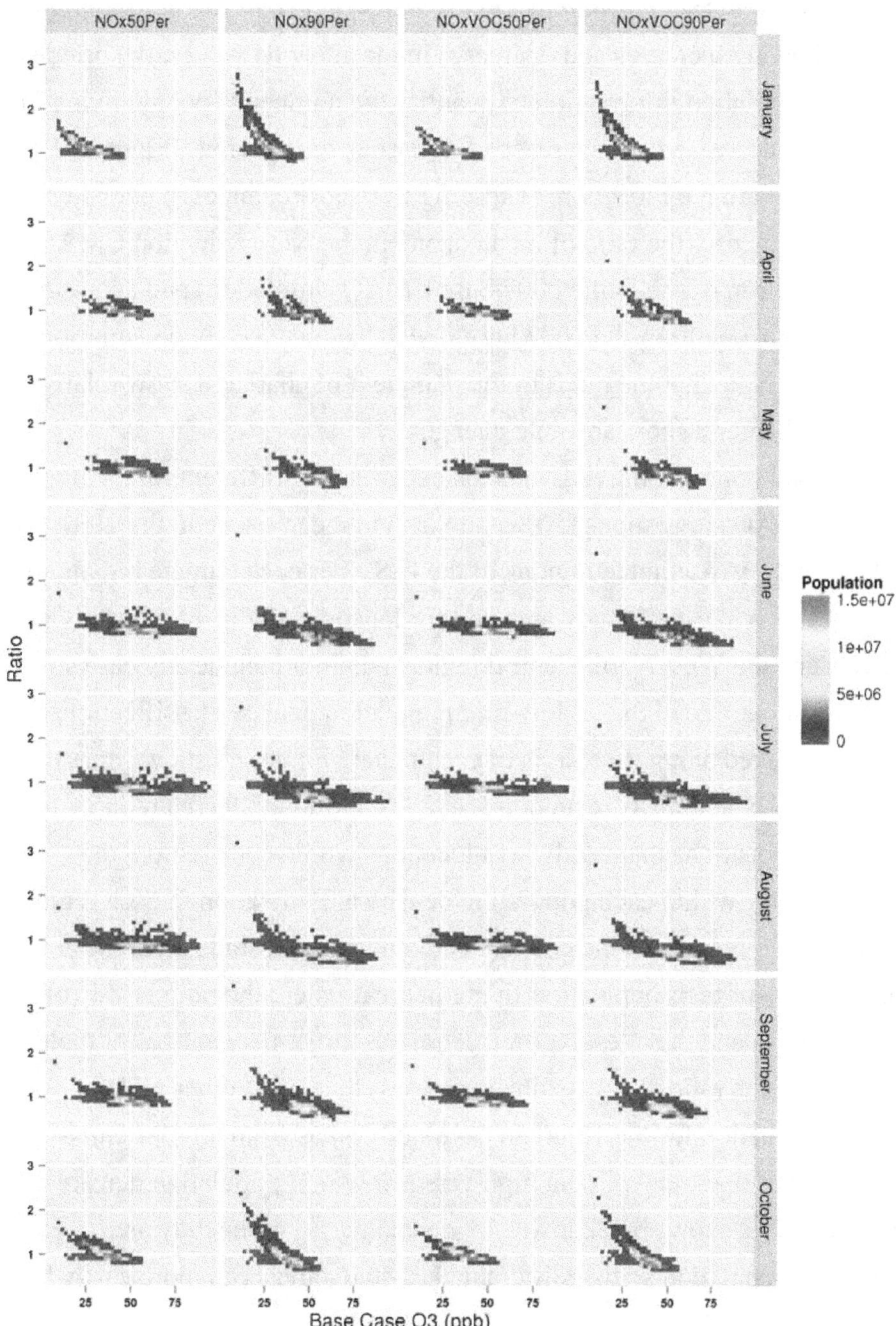

Figure 49: Density scatter plot comparing modeled monthly mean ozone in the 2007 base CMAQ simulation to the relative change in monthly mean ozone from the emissions reduction CMAQ simulations. Relative change is shown as the ratio of ozone in the emissions reduction simulation to ozone in the 2007 base simulation. Colors depict the number of people living in areas that fall at any location on the scatter plot.

2.2 MODELED OZONE RESPONSE PAIRED WITH POPULATION

In addition to maps showing increases and decreases in mean ozone values, the gridded model data were paired with population information to quantify the number of people living in locations where modeled ozone decreased and increased for various time periods. Figure 50-Figure 60 break down this information by location. These figures show changes in ozone using two different metrics: a relative metric (the ratio of mean ozone in the NOx reduction CMAQ simulations (50% and 90%) to mean ozone in the 2007 base CMAQ simulation) and an absolute metric (the ppb change in mean ozone from the 2007 base CMAQ simulation to the emissions reduction CMAQ simulations). Note that the maps in the main text of chapter 8 show relative changes while the barplots in chapter 8 show absolute changes.

Figure 50 shows the total population living in areas experiencing different ratios of mean ozone in the NOx reduction CMAQ simulations (50% and 90%) to mean ozone in the 2007 base CMAQ simulation for the nine NOAA climate regions of the U.S. For each climate region, this information is shown for locations in a case-study area and for locations not in a case-study area. Two regions, the Northwest and the West North Central regions, did not include any case study areas. Each area is further split out into high and low-mid population density classifications. Values for each month are displayed along the x-axis of each panel. Figure 51 shows the same information for the combined NOx and VOC reduction scenarios. Although there are more total people living in non-study area locations than study area locations within each region, the patterns in the two look similar for within each population density classification in each region. It should be noted that for the two regions of the country without an urban study area, the Northwest has larger percentages of their population living in areas where the ratio is > 1 (ozone increases) than most other regions and the West North Central has larger percentages of their population living in areas where the ratio is < 1 (ozone decreases) than most other regions. Figure 52 shows the same information for the 15 urban case study areas from all four emissions reduction CMAQ simulations but does not split out high versus low-mid population density locations. Also note that Figure 52 shows breakdowns by percentage of case-study area population rather than by total population so that different case-study areas can more easily be compared.

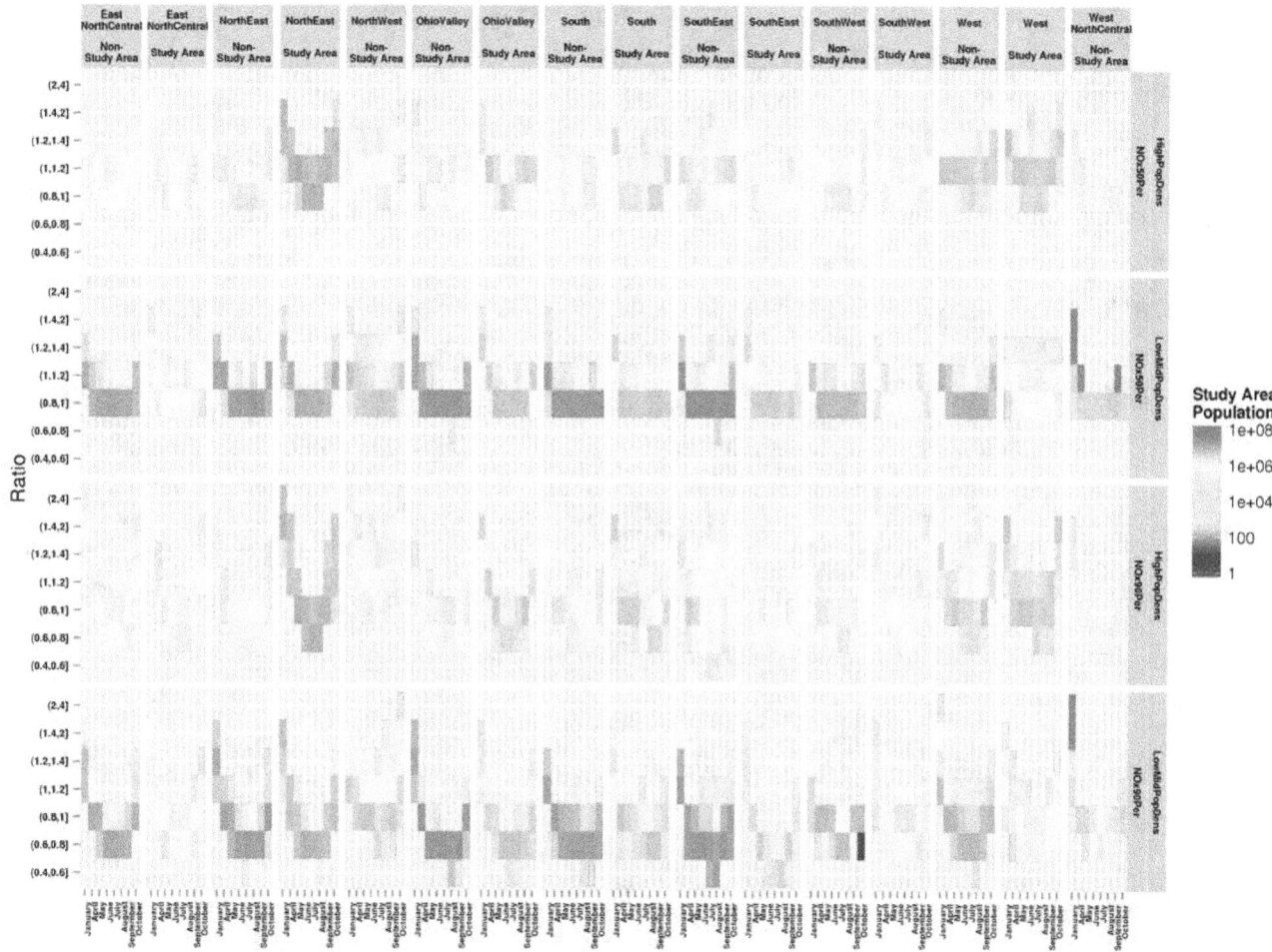

Figure 50: P Populations living in locations with various ranges of ratios of monthly mean ozone in the NOx reduction simulations to monthly mean ozone in the 2007 base CMAQ simulation. Eight different monthly ratios are shown in each panel (January, April-October). Panels split population by 9 climate regions, urban case study area vs non-urban case study area, urban versus non-urban and 50% NOx reduction scenario vs 90% NOx reduction scenario.

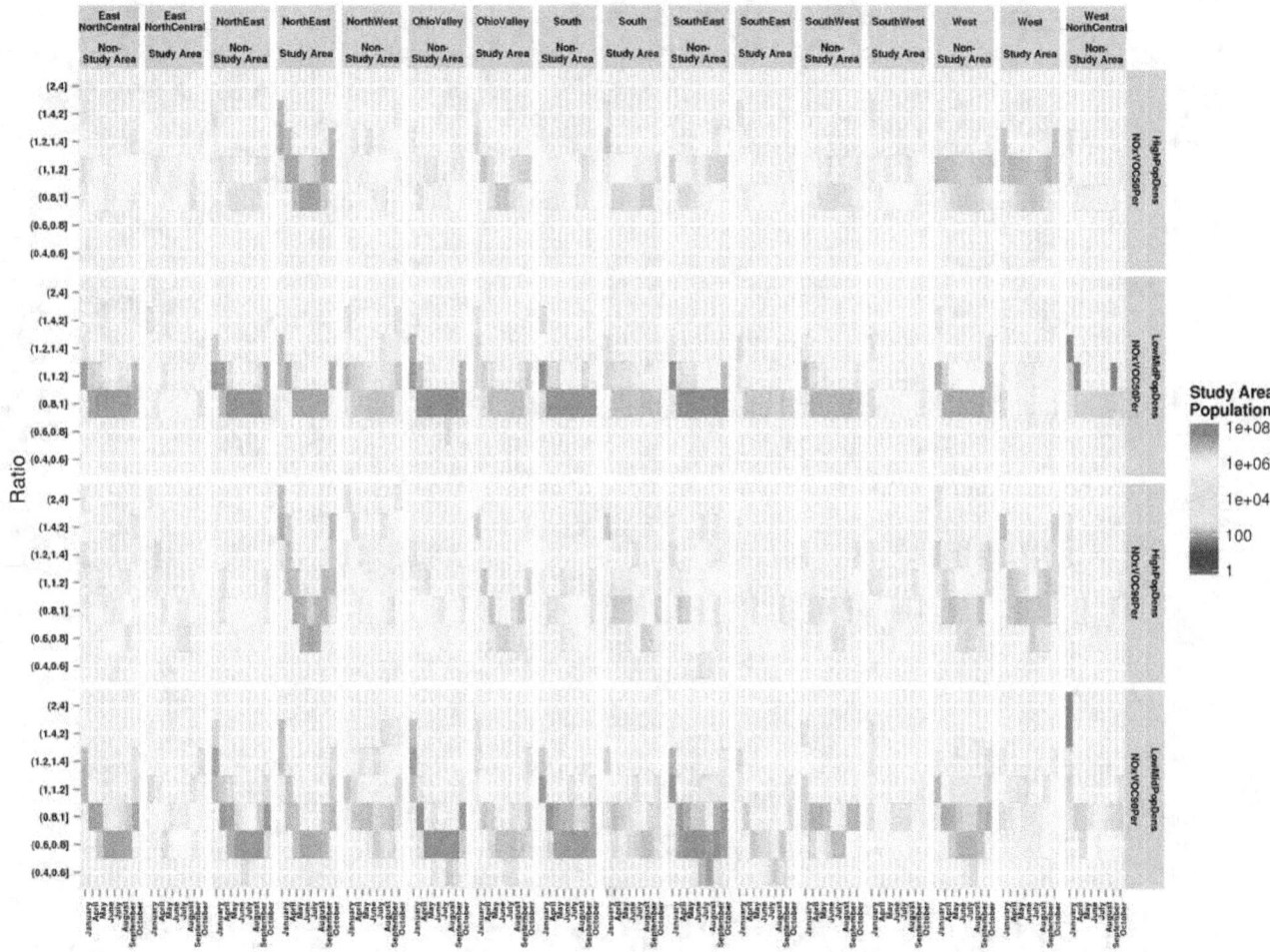

Figure 51: Populations living in locations with various ranges of ratios of monthly mean ozone in the combined NOx and VOC reduction simulations to monthly mean ozone in the 2007 base CMAQ simulation. Eight different monthly ratios are shown in each panel (January, April-October). Panels split population by 9 climate regions, urban case study area vs non-urban case study area, urban versus non-urban and 50% NOx/VOC reduction scenario vs 90% NOx/VOC reduction scenario.

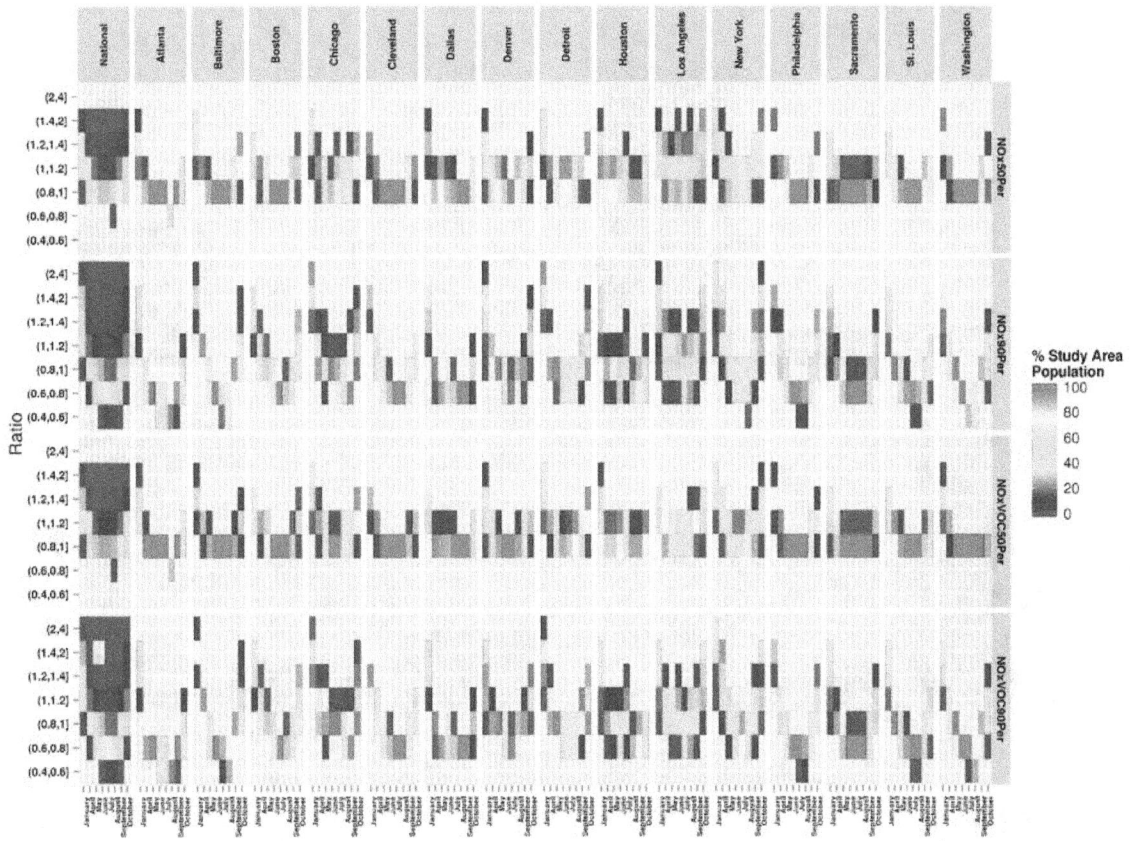

Figure 52: Percent of population living in locations with various ranges of ratios of monthly mean ozone in the four emissions reduction simulations to monthly mean ozone in the 2007 base CMAQ simulation. Eight different monthly ratios are shown in each panel (January, April-October). Panels split population by 15 urban case-study areas and by four emissions reduction simulations: from top to bottom, 50% NOx reduction, 90% NOx reduction, 50% NOx and VOC reduction, 90% NOx and VOC reduction.

Section 8.2.3 further examined these ozone ratios using histograms and lumping all study areas together and all non-study areas together. The main text provided histograms for the NOx reduction scenarios only. This appendix provides histograms for all four emission reduction simulations in using both relative and absolute metrics (Figure 53-Figure 60). These figures show that the breakdown of people living in locations of increasing versus decreasing ozone for various monthly and seasonal time-periods does not change much between the NOx reduction scenarios and the equivalent NOx and VOC reduction scenarios. Table 1 provides the numbers going into the 50% NOx reduction and 90% NOx reduction histograms. Table 2 and Table 3 break down the April-October seasonal mean ozone results by two further classification schemes: high versus low-mid population density locations and by the 15 case study areas.

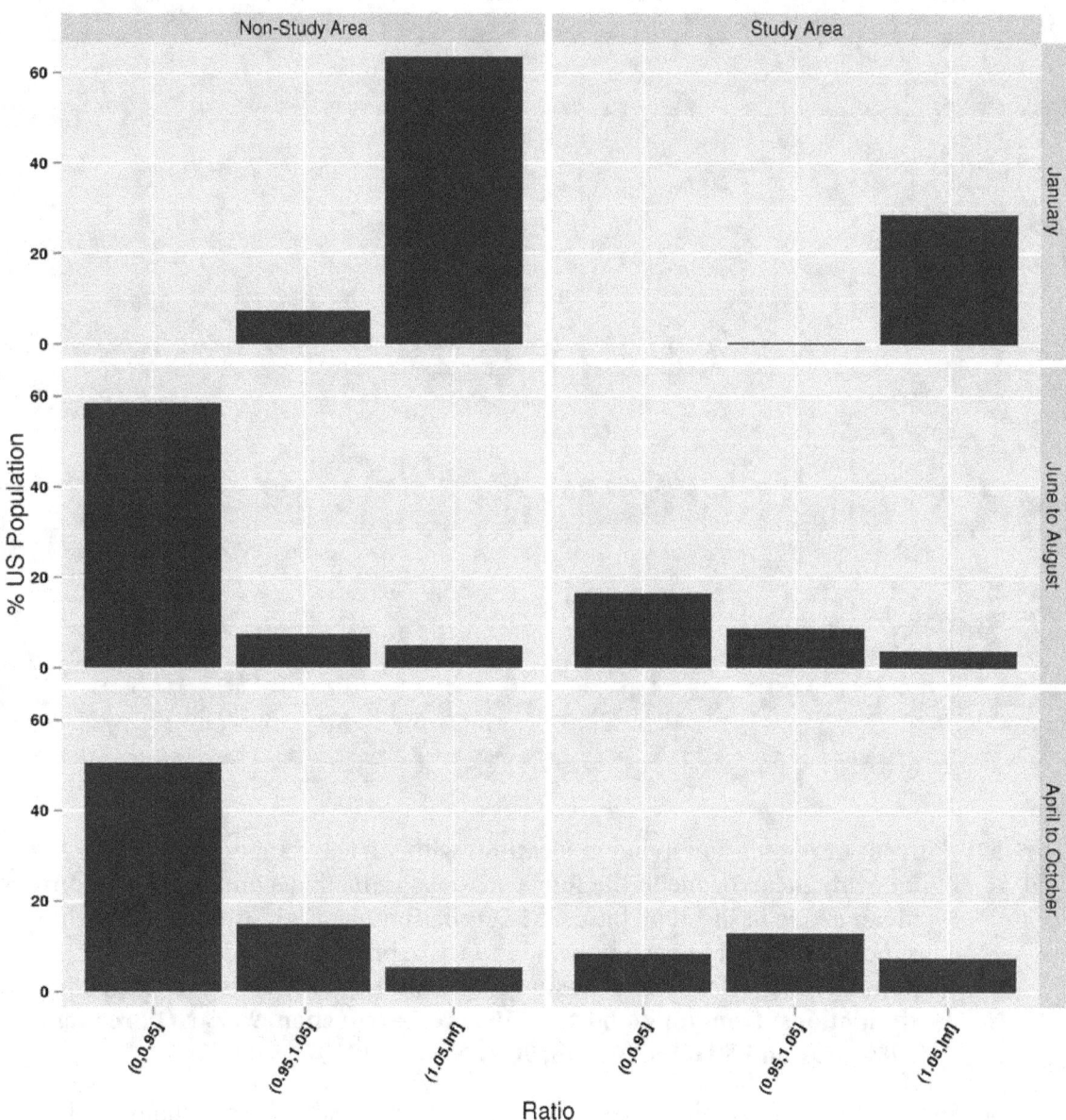

Figure 53: Histograms of US population living in locations with increasing and decreasing mean ozone. Values on the x-axis represent the ratio of mean ozone in the 50% NOx cut CMAQ simulation to the mean ozone in the 2007 base CMAQ simulation. The percentages of the US population living in areas that have ratios less than 0.95, from 0.95 to 1.05 and greater than 1.05 are shown on the y-axis. Left plots show population numbers in locations not included in one of the urban case study areas while right plots show population numbers in locations included in one of the urban case study areas. Top plots show ratios of January monthly mean ozone, middle plots show ratios of season mean June-August ozone, and bottom plots show ratios of seasonal mean April-October ozone.

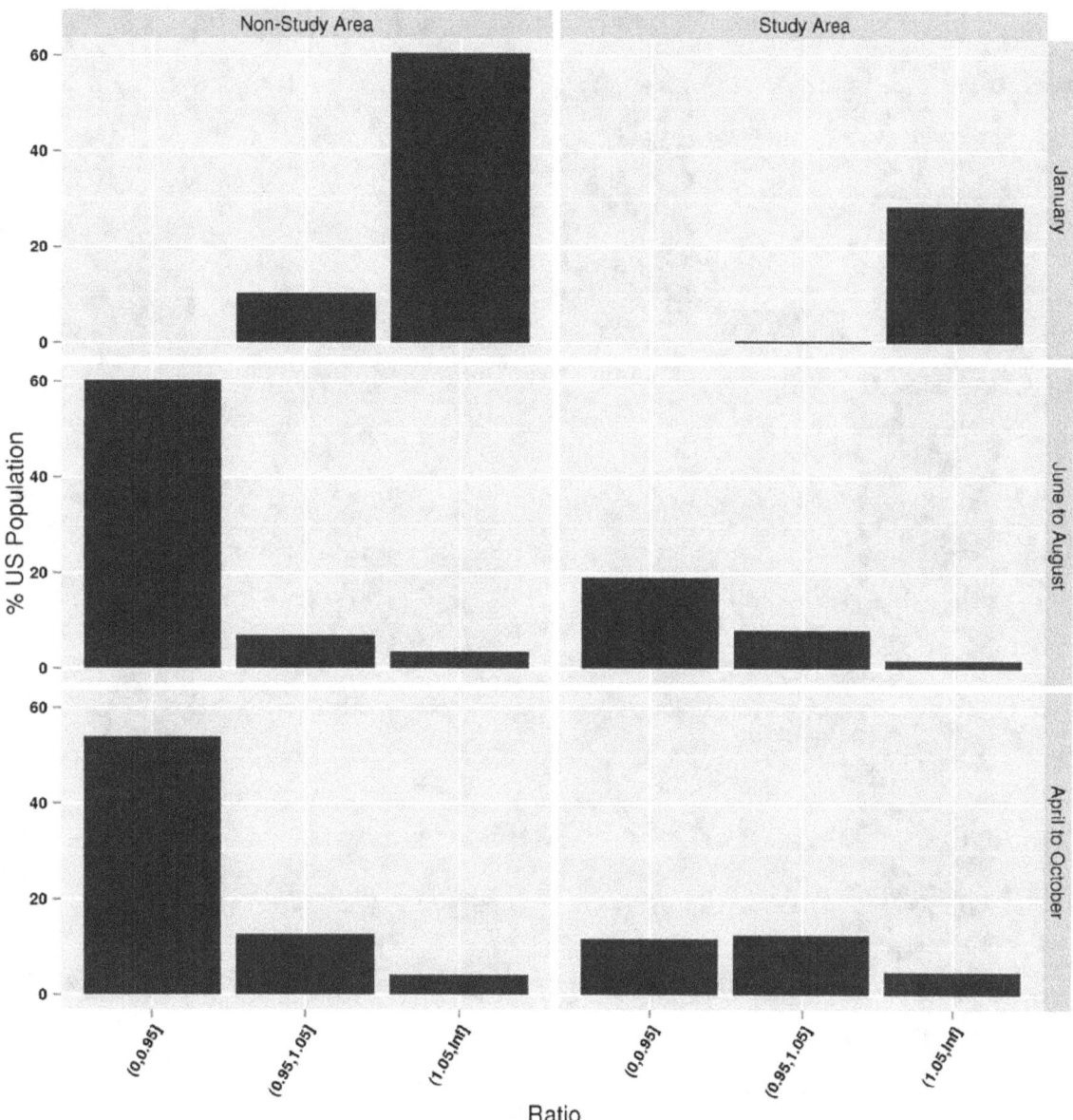

Figure 54: Histograms of US population living in locations with increasing and decreasing mean ozone. Values on the x-axis represent the ratio of mean ozone in the 50% NOx and VOC cut CMAQ simulation to the mean ozone in the 2007 base CMAQ simulation. The percentages of the US population living in areas that have ratios less than 0.95, from 0.95 to 1.05 and greater than 1.05 are shown on the y-axis. Left plots show population numbers in locations not included in one of the urban case study areas while right plots show population numbers in locations included in one of the urban case study areas. Top plots show ratios of January monthly mean ozone, middle plots show ratios of season mean June-August ozone, and bottom plots show ratios of seasonal mean April-October ozone.

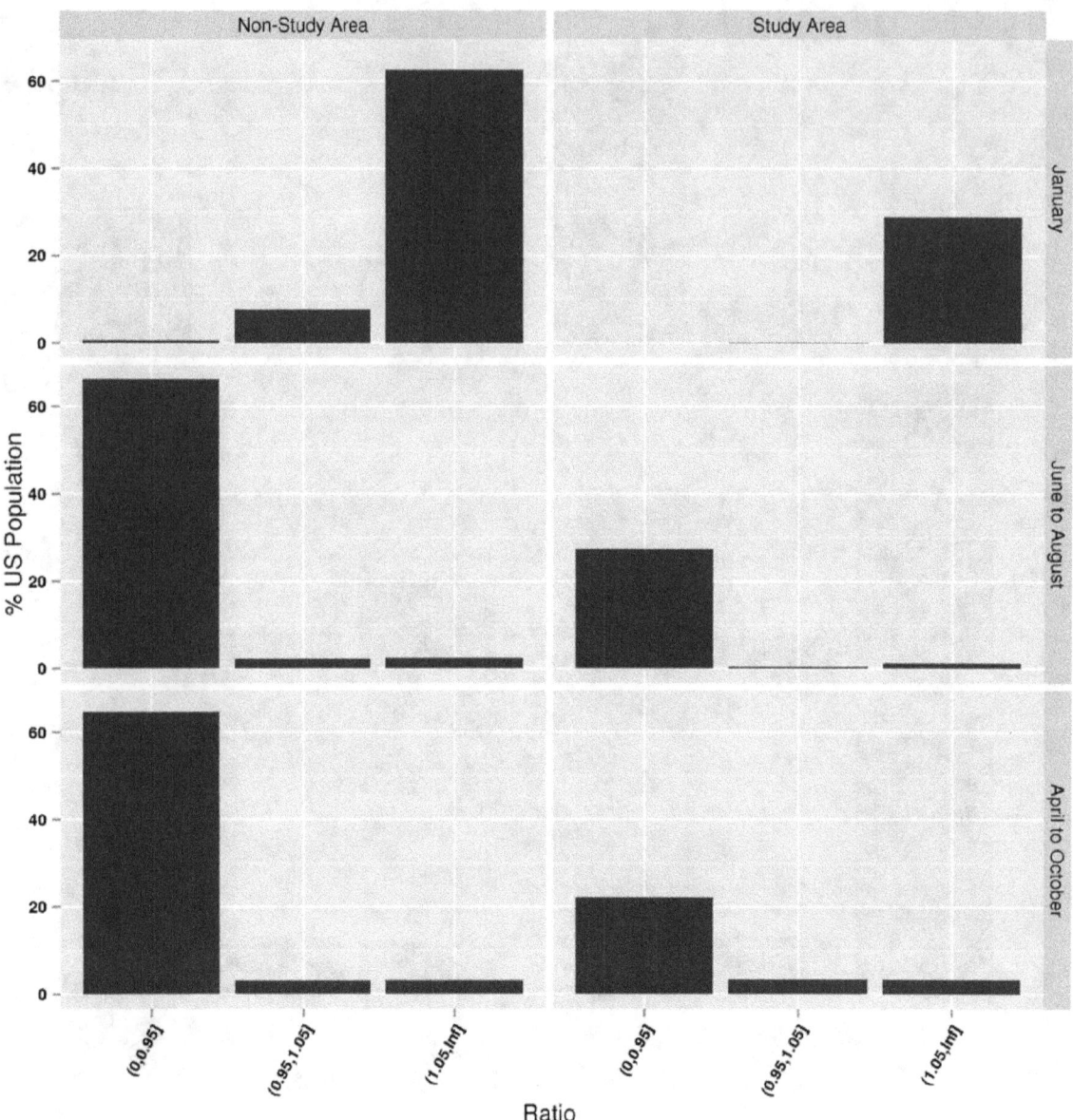

Figure 55: Histograms of US population living in locations with increasing and decreasing mean ozone. Values on the x-axis represent the ratio of mean ozone in the 90% NOx cut CMAQ simulation to the mean ozone in the 2007 base CMAQ simulation. The percentages of the US population living in areas that have ratios less than 0.95, from 0.95 to 1.05 and greater than 1.05 are shown on the y-axis. Left plots show population numbers in locations not included in one of the urban case study areas while right plots show population numbers in locations included in one of the urban case study areas. Top plots show ratios of January monthly mean ozone, middle plots show ratios of season mean June-August ozone, and bottom plots show ratios of seasonal mean April-October ozone.

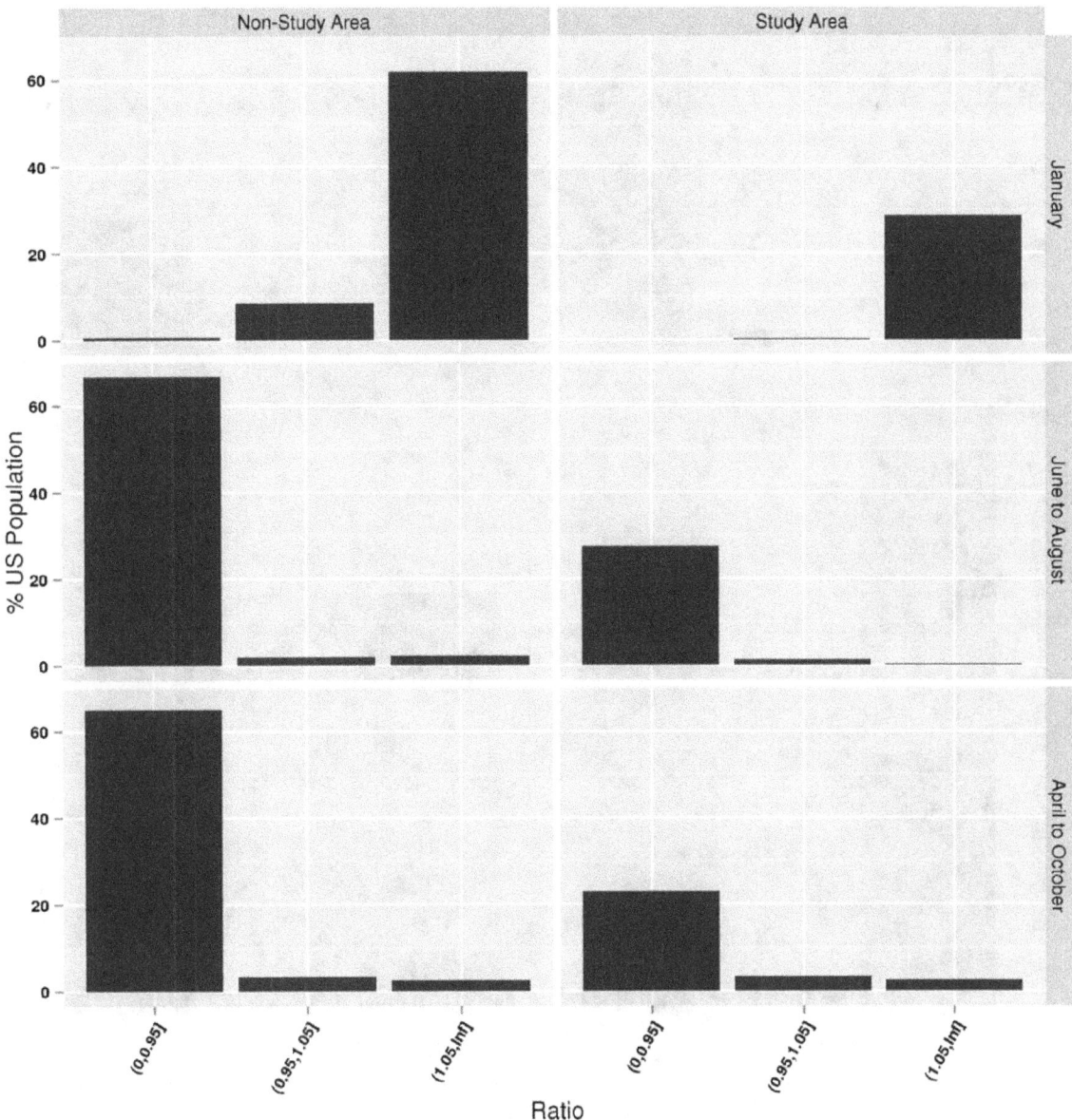

Figure 56: Histograms of US population living in locations with increasing and decreasing mean ozone. Values on the x-axis represent the ratio of mean ozone in the 90% NOx and VOC cut CMAQ simulation to the mean ozone in the 2007 base CMAQ simulation. The percentages of the US population living in areas that have ratios less than 0.95, from 0.95 to 1.05 and greater than 1.05 are shown on the y-axis. Left plots show population numbers in locations not included in one of the urban case study areas while right plots show population numbers in locations included in one of the urban case study areas. Top plots show ratios of January monthly mean ozone, middle plots show ratios of season mean June-August ozone, and bottom plots show ratios of seasonal mean April-October ozone.

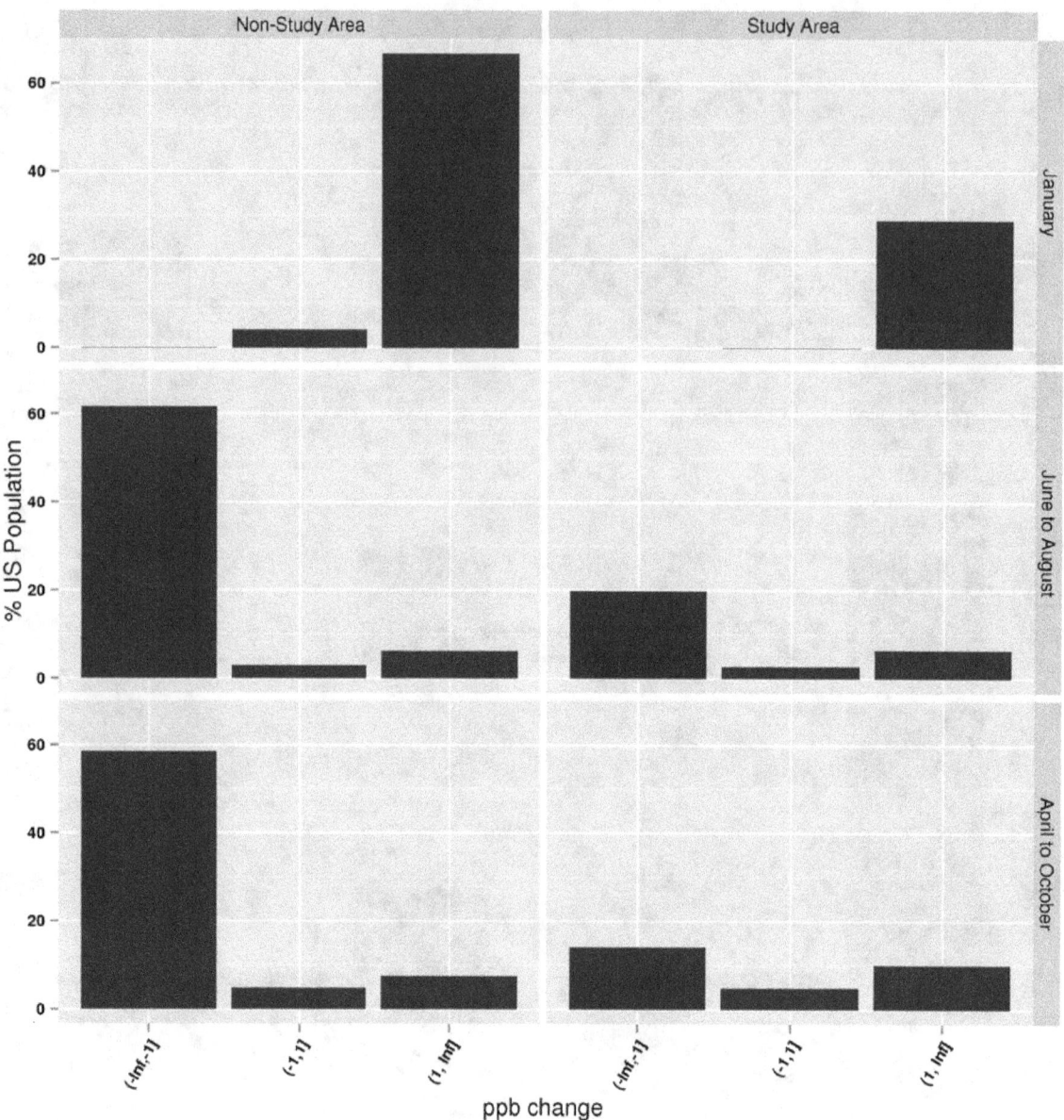

Figure 57: Histograms of US population living in locations with increasing and decreasing mean ozone. Values on the x-axis represent the absolute (ppb) change of mean ozone from the 2007 base CMAQ simulation to the 50% NOx cut CMAQ simulation to the mean ozone in the 2007 base CMAQ simulation. The percentages of the US population living in areas that have changes less than -1 ppb, between -1 and +1 ppb and greater than +1 ppb are shown on the y-axis. Left plots show population numbers in locations not included in one of the case study areas while right plots show population numbers in locations included in one of the case study areas. Top plots show ratios of January monthly mean ozone, middle plots show ratios of season mean June-August ozone, and bottom plots show ratios of seasonal mean April-October ozone.

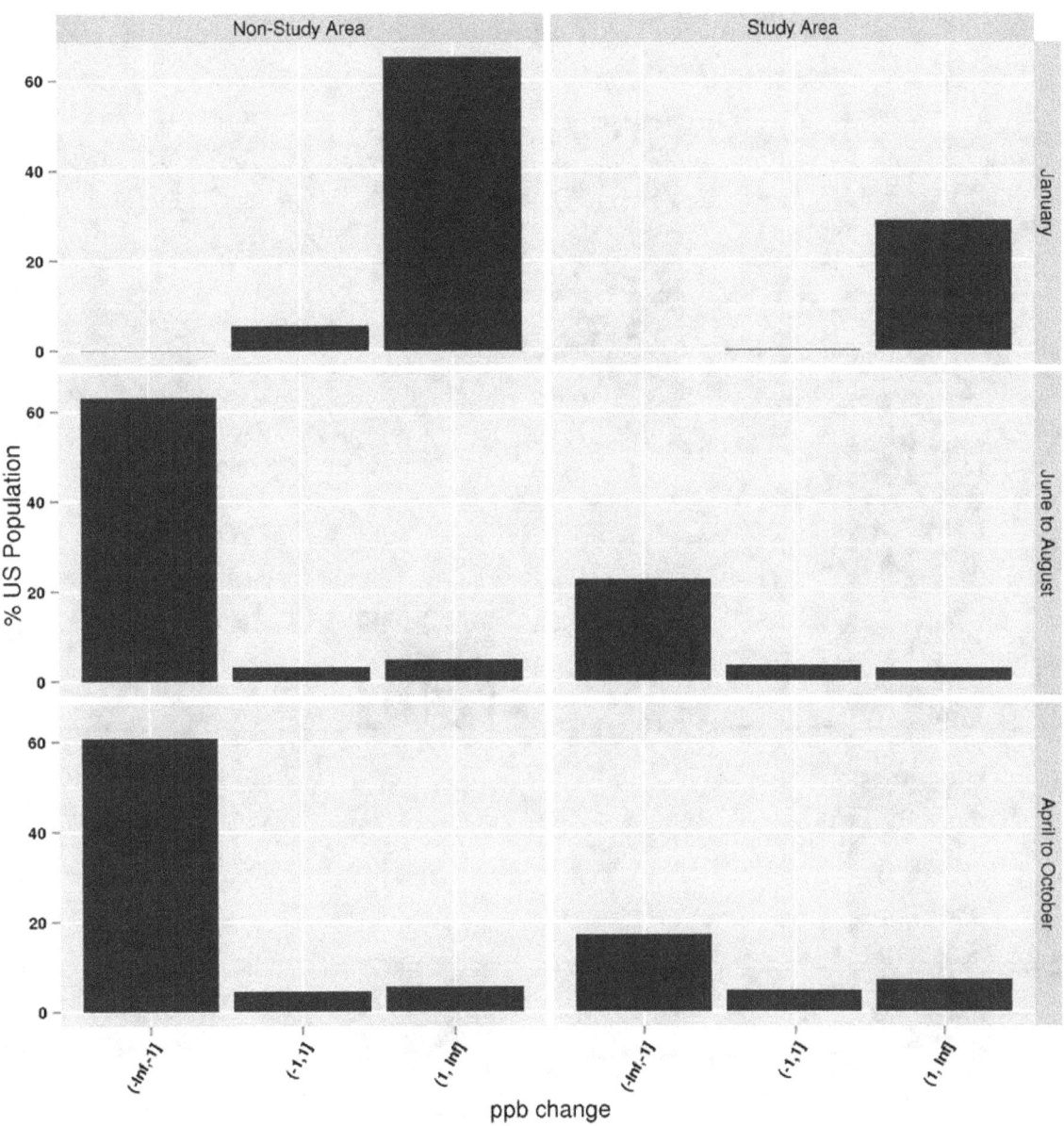

Figure 58: Histograms of US population living in locations with increasing and decreasing mean ozone. Values on the x-axis represent the absolute (ppb) change of mean ozone from the 2007 base CMAQ simulation to the 50% NOx and VOC cut CMAQ simulation to the mean ozone in the 2007 base CMAQ simulation. The percentages of the US population living in areas that have changes less than -1 ppb, between -1 and +1 ppb and greater than +1 ppb are

shown on the y-axis. Left plots show population numbers in locations not included in one of the case study areas while right plots show population numbers in locations included in one of the case study areas. Top plots show ratios of January monthly mean ozone, middle plots show ratios of season mean June-August ozone, and bottom plots show ratios of seasonal mean April-October ozone.

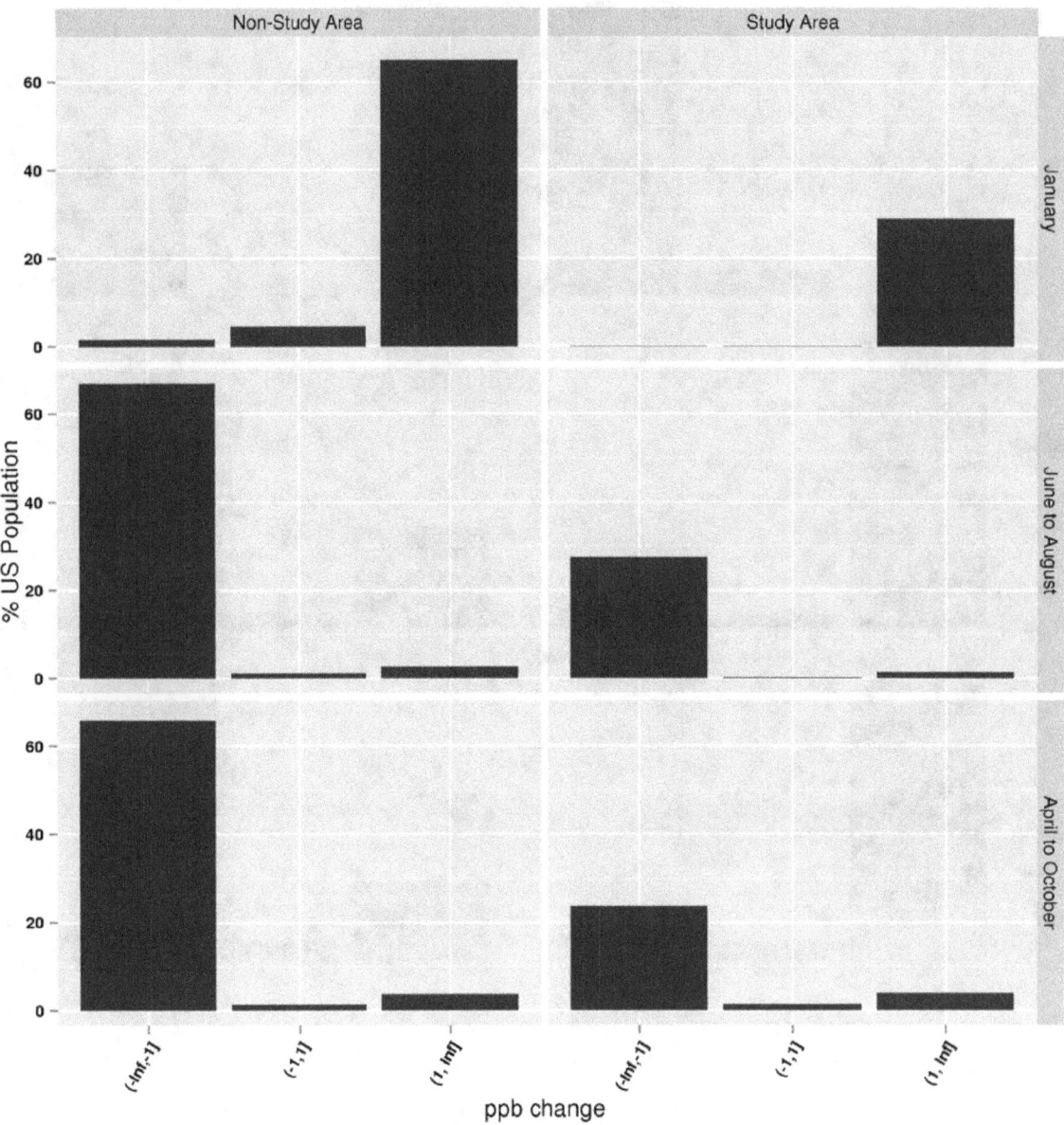

Figure 59: Histograms of US population living in locations with increasing and decreasing mean ozone. Values on the x-axis represent the absolute (ppb) change of mean ozone from the 2007 base CMAQ simulation to the 90% NOx cut CMAQ simulation to the mean ozone in the 2007 base CMAQ simulation. The percentages of the US population living in areas that have changes less than -1 ppb, between -1 and +1 ppb and greater than +1 ppb are shown on

the y-axis. Left plots show population numbers in locations not included in one of the case study areas while right plots show population numbers in locations included in one of the case study areas. Top plots show ratios of January monthly mean ozone, middle plots show ratios of season mean June-August ozone, and bottom plots show ratios of seasonal mean April-October ozone.

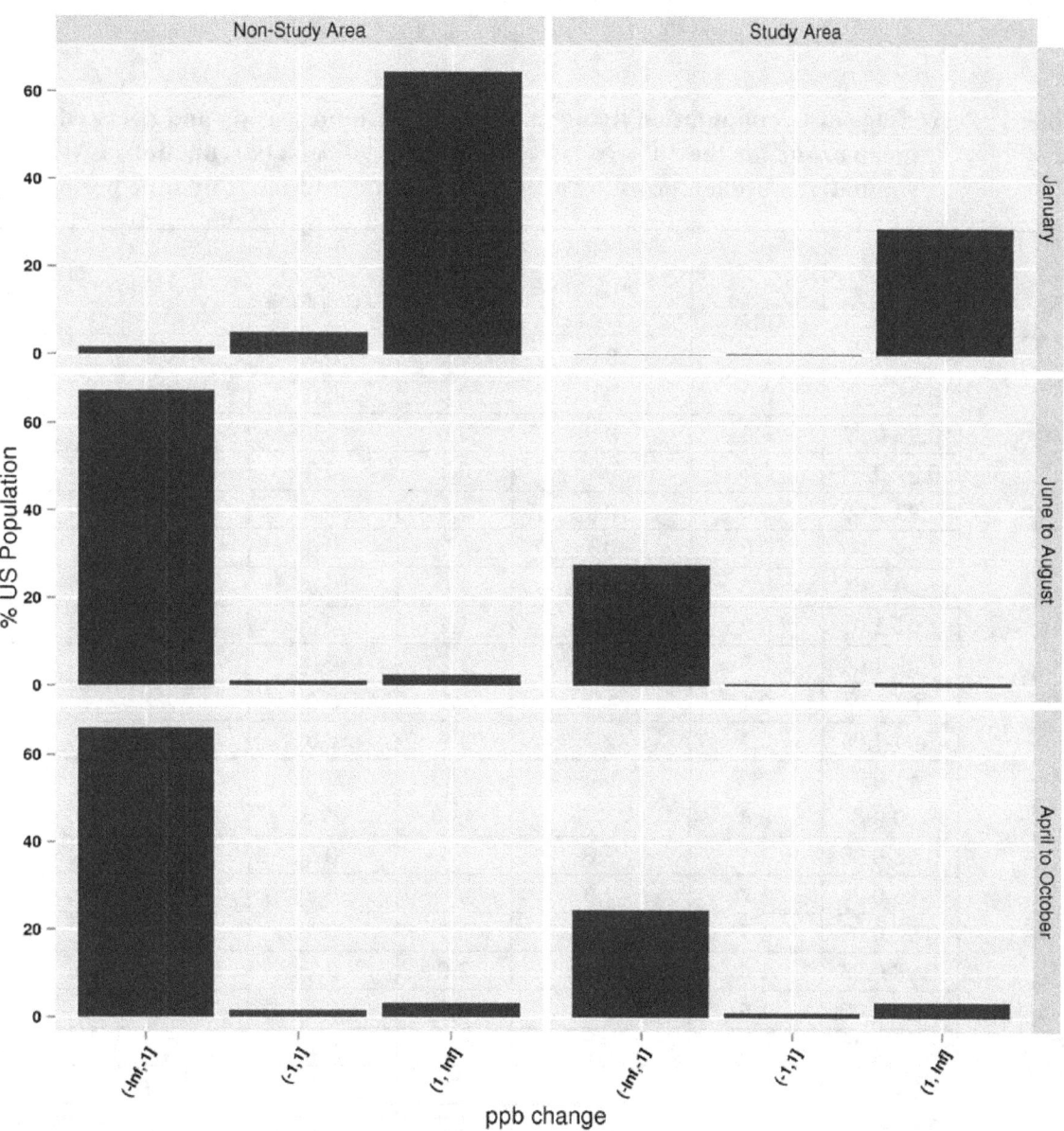

Figure 60: Histograms of US population living in locations with increasing and decreasing mean ozone. Values on the x-axis represent the absolute (ppb) change of mean ozone from the 2007 base CMAQ simulation to the 90% NOx and VOC cut CMAQ simulation to the mean ozone in the 2007 base CMAQ

simulation. The percentages of the US population living in areas that have changes less than -1 ppb, between -1 and +1 ppb and greater than +1 ppb are shown on the y-axis. Left plots show population numbers in locations not included in one of the case study areas while right plots show population numbers in locations included in one of the case study areas. Top plots show ratios of January monthly mean ozone, middle plots show ratios of season mean June-August ozone, and bottom plots show ratios of seasonal mean April-October ozone.

Table 1: Percentage of US population living in locations with increasing and decreasing mean ozone for the 50% NOx reduction and 90% NOx reductions CMAQ simulations broken down by different seasonal and monthly time periods.

	Ratio	50% NOx			90% NOx		
		Study Area	Non Study Area	US	Study Area	Non Study Area	US
January	<0.95	0.0	0.0	0.0	0.0	0.6	0.6
	0.95-0.96	0.0	0.0	0.0	0.0	0.3	0.3
	0.96-0.97	0.0	0.0	0.0	0.0	0.3	0.3
	0.97-0.98	0.1	0.1	0.1	0.0	0.5	0.5
	0.98-0.99	0.0	0.4	0.4	0.0	0.7	0.7
	0.99-1.00	0.0	0.7	0.7	0.0	0.7	0.8
	1.00-1.01	0.0	0.9	1.0	0.0	0.8	0.8
	1.01-1.02	0.0	1.1	1.1	0.0	0.8	0.8
	1.02-1.03	0.0	1.1	1.2	0.0	1.0	1.1
	1.03-1.04	0.1	1.4	1.5	0.0	1.2	1.2
	1.04-1.05	0.2	1.7	1.9	0.0	1.3	1.3
	>1.05	28.7	63.6	92.3	28.9	62.6	91.5
April-October	<0.95	8.4	50.6	59.0	22.3	64.7	87.0
	0.95-0.96	2.1	3.6	5.7	0.5	0.2	0.8
	0.96-0.97	1.7	2.8	4.5	0.6	0.6	1.2
	0.97-0.98	1.7	1.9	3.7	0.2	0.5	0.7
	0.98-0.99	1.4	1.6	3.0	0.5	0.2	0.6
	0.99-1.00	1.5	1.4	2.9	0.5	0.2	0.8
	1.00-1.01	0.8	0.7	1.5	0.2	0.5	0.6
	1.01-1.02	1.1	0.7	1.8	0.2	0.2	0.3
	1.02-1.03	0.8	0.9	1.7	0.2	0.3	0.5
	1.03-1.04	1.1	0.5	1.6	0.4	0.3	0.7
	1.04-1.05	0.7	0.7	1.4	0.1	0.2	0.3
	>1.05	7.6	5.5	13.1	3.4	3.2	6.6
June-	<0.95	16.4	58.4	74.8	27.4	66.3	93.7

August	Ratio	Study Area	Non Study Area	US	Study Area	Non Study Area	US
August	0.95-0.96	1.1	1.6	2.7	0.1	0.3	0.4
	0.96-0.97	1.1	0.9	1.9	0.0	0.3	0.3
	0.97-0.98	1.1	1.1	2.1	0.0	0.3	0.4
	0.98-0.99	1.0	0.9	1.8	0.0	0.2	0.2
	0.99-1.00	1.1	0.7	1.8	0.0	0.2	0.2
	1.00-1.01	0.5	0.6	1.1	0.1	0.3	0.4
	1.01-1.02	0.7	0.5	1.2	0.1	0.2	0.3
	1.02-1.03	0.5	0.5	0.9	0.0	0.0	0.1
	1.03-1.04	0.6	0.4	1.0	0.0	0.2	0.2
	1.04-1.05	1.3	0.4	1.7	0.1	0.0	0.2
	>1.05	3.8	5.0	8.8	1.2	2.4	3.7

Table 2: Percentage of US population living in locations with increasing and decreasing April-October seasonal mean ozone in the 50% NOx reduction and 90% NOx reductions CMAQ simulations broken down by high and low-mid population density areas.

	Ratio	50% NOx			90% NOx		
		Study Area	Non Study Area	US	Study Area	Non Study Area	US
High population density	<0.95	0.8	1.7	2.5	9.8	6.7	16.5
	0.95-0.96	0.6	0.7	1.3	0.5	0.1	0.6
	0.96-0.97	0.8	0.6	1.4	0.6	0.4	1.0
	0.97-0.98	0.9	0.5	1.5	0.2	0.3	0.5
	0.98-0.99	0.9	0.6	1.6	0.5	0.1	0.6
	0.99-1.00	1.1	0.7	1.8	0.5	0.1	0.6
	1.00-1.01	0.7	0.3	1.0	0.2	0.3	0.5
	1.01-1.02	0.9	0.4	1.3	0.2	0.1	0.2
	1.02-1.03	0.7	0.7	1.4	0.2	0.1	0.3
	1.03-1.04	1.0	0.3	1.3	0.4	0.2	0.6
	1.04-1.05	0.7	0.5	1.1	0.1	0.1	0.1
	>1.05	7.3	3.8	11.1	3.4	2.3	5.7
Low-Mid population density	<0.95	7.6	48.9	56.5	12.5	58.1	70.6
	0.95-0.96	1.5	3.0	4.5	0.1	0.2	0.2
	0.96-0.97	0.9	2.1	3.1	0.0	0.1	0.1
	0.97-0.98	0.8	1.4	2.2	0.0	0.2	0.2
	0.98-0.99	0.5	0.9	1.4	0.0	0.1	0.1
	0.99-1.00	0.4	0.7	1.1	0.0	0.1	0.1
	1.00-1.01	0.1	0.4	0.5	0.0	0.1	0.2
	1.01-1.02	0.2	0.3	0.5	0.0	0.1	0.1
	1.02-1.03	0.1	0.3	0.3	0.0	0.1	0.1
	1.03-1.04	0.1	0.2	0.3	0.0	0.1	0.1
	1.04-1.05	0.0	0.2	0.3	0.0	0.1	0.1

>1.05	0.3	1.7	2.0	0.0	0.8	0.9

Table 3: Percentage of US population living in locations with increasing and decreasing April-October seasonal mean ozone in the 50% NOx reduction and 90% NOx reduction CMAQ simulations broken down by 15 case study areas.

Scenario	Study Area	Ratio of April-October seasonal mean ozone in reduced emissions CMAQ simulation to April-October seasonal mean ozone in base 2007 CMAQ simulation											
		0-0.95	0.95-0.96	0.96-0.97	0.97-0.98	0.98-0.99	0.99-1.00	1.00-1.01	1.01-1.02	1.02-1.03	1.03-.04	1.04-1.05	>1.05
50% NOx reduction	Not in Study Area	50.6	3.6	2.8	1.9	1.6	1.4	0.7	0.7	0.9	0.5	0.7	5.5
	Atlanta	1.6	0.1	0.0	0.0	0.0	0.0	0.0	0.0	0.0	0.0	0.0	0.0
	Baltimore	0.4	0.0	0.2	0.0	0.1	0.0	0.0	0.0	0.0	0.0	0.1	0.0
	Boston	0.4	0.2	0.1	0.1	0.1	0.1	0.0	0.2	0.0	0.0	0.2	0.0
	Chicago	0.5	0.1	0.2	0.2	0.3	0.1	0.1	0.2	0.0	0.2	0.1	0.9
	Cleveland	0.2	0.1	0.1	0.0	0.1	0.1	0.0	0.1	0.0	0.0	0.0	0.0
	Dallas	1.0	0.4	0.2	0.1	0.1	0.1	0.0	0.1	0.0	0.0	0.0	0.0
	Denver	0.1	0.2	0.1	0.0	0.1	0.1	0.0	0.1	0.1	0.0	0.0	0.1
	Detroit	0.1	0.1	0.1	0.0	0.1	0.1	0.0	0.1	0.2	0.2	0.1	0.3
	Houston	0.8	0.1	0.0	0.2	0.1	0.1	0.1	0.0	0.1	0.1	0.1	0.2
	Los Angeles	0.2	0.1	0.1	0.0	0.0	0.4	0.0	0.1		0.2	0.0	2.8
	New York	0.5	0.3	0.3	0.2	0.2	0.3	0.1	0.2	0.2	0.1	0.1	3.4
	Philadelphia	0.7	0.1	0.2	0.3	0.2	0.2	0.1	0.0	0.2	0.2	0.0	0.0
	Sacramento	0.2	0.2	0.0	0.1	0.1	0.0	0.1	0.0	0.0	0.0	0.0	0.0
	St. Louis	0.6	0.1	0.0	0.1	0.0	0.0	0.1	0.0	0.0	0.0	0.0	0.0
	Washington	1.1	0.1	0.2	0.2	0.1	0.0	0.1	0.0	0.0	0.0	0.0	0.0
90% NOx reduction	Not in Study Area	64.7	0.2	0.6	0.5	0.2	0.2	0.5	0.2	0.3	0.3	0.2	3.2
	Atlanta	1.7	0.0	0.0	0.0	0.0	0.0	0.0	0.0	0.0	0.0	0.0	0.0
	Baltimore	0.9	0.0	0.0	0.0	0.0	0.0	0.0	0.0	0.0	0.0	0.0	0.0
	Boston	1.2	0.0	0.2	0.0	0.0	0.0	0.0	0.0	0.0	0.0	0.0	0.0
	Chicago	2.4	0.0	0.3	0.0	0.1	0.3	0.0	0.0	0.0	0.0	0.0	0.1
	Cleveland	0.6	0.0	0.0	0.0	0.0	0.0	0.0	0.0	0.0	0.0	0.0	0.0
	Dallas	2.0	0.0	0.0	0.0	0.0	0.0	0.0	0.0	0.0	0.0	0.0	0.0
	Denver	0.7	0.0	0.0	0.1	0.0	0.0	0.0	0.0	0.0	0.0	0.0	0.0
	Detroit	1.0	0.1	0.0	0.0	0.1	0.0	0.1	0.1	0.1	0.0	0.0	0.0
	Houston	1.7	0.0	0.0	0.0	0.0	0.0	0.0	0.0	0.1	0.0	0.1	0.0
	Los Angeles	1.7	0.1	0.2	0.1	0.0	0.2	0.1	0.1	0.0	0.2	0.0	1.4
	New York	3.2	0.3	0.0	0.0	0.3	0.1	0.0	0.0	0.0	0.2	0.0	1.9
	Philadelphia	2.0	0.0	0.0	0.0	0.0	0.0	0.0	0.0	0.0	0.0	0.0	0.0

	Sacramento	0.7	0.0	0.0	0.0	0.0	0.0	0.0	0.0	0.0	0.0	0.0	0.0
	St. Louis	0.9	0.0	0.0	0.0	0.0	0.0	0.0	0.0	0.0	0.0	0.0	0.0
	Washington	1.8	0.0	0.0	0.0	0.0	0.0	0.0	0.0	0.0	0.0	0.0	0.0

Appendix 9-A
Exposure and Lung-Function Risk Estimates for Sub-Regions of Each Study Area (urban core, outer ring, total regions)

Simulated populations within different sub-regions of a given urban study area may have different exposure and lung-function risk distributions reflecting potential differences both in their patterns of behavior (e.g., commuting patterns, outdoor activities) as well as differences in the spatio-temporal ambient ozone fields estimated for each sub-region. To explore potential spatial heterogeneity in both the exposure and lung-function risk estimates, we have completed a stratified analysis of risk for both of these assessments. These stratified analysis consider two sub-regions within each study area including: (a) a smaller urban core sub-area matching that used in the Smith et al., 2009 epidemiology study providing the effect estimates used in modeling short-term exposure-based mortality risk and (b) the outer ring reflecting the remainder of the larger study area used in the exposure and lung-function assessment (excluding the core urban area). In presenting risk estimates based on these two sub-regions, we also include risk estimates based on the entire study area for completeness.

Generating these sub-region risk estimates is relatively straight-forward. As part of our standard APEX output for the exposure and lung function risk estimates summarized in Chapters 5 and 6 respectively, limited exposure and FEV1 results are retained for each simulated person including their daily maximum 1-hour ozone exposure and counts per ozone season of each time a simulated person experienced an FEV1 decrement (10%, 15%, and 20%). Also retained is the location of their home census tract (and corresponding location of ambient concentration source used for calculating exposures) within the larger study areas used in the exposure and lung-function analyses. To generate the sub-region estimates, we subset these broader study area exposure and FEV1 risk results into two sets of exposure results for each of 12 study areas: one containing those persons residing within the urban core and the other containing persons residing in the outer ring outside the urban core. In addition, two years of data were evaluated for the 12 study areas (2007 and 2009), matching the two years for which short term mortality risks were estimated. In generating these sub-region estimates, we focused on the 12 urban study areas used in the epidemiology-based risk assessment to allow these stratified results to be compared alongside the urban core and CBSA-based estimates generated as part of the epidemiological-based risk assessment.

In summarizing these risk estimates, we first focus on the exposure estimates (figures 9A-1 through 9A-12), including the percent of all simulated individuals experiencing 1-hour exposure at or above each specified benchmark (see Chapter 5 for additional detail on this risk metric). Estimates are presented for both 2007 and 2009 within each figure. In order to compare, for example, exposure estimates (based on the 60 ppb benchmark) for the urban core between current conditions and the current standard for 2007, we would compare the darker blue column for *urb_122_base_07* with the darker blue column for *urb_122_75_07*.

After presenting the exposure estimates, we then present lung-function estimates (figures 9A-13 through 9A-24) including percent of all simulated individuals experiencing at least one FEV1 decrement of 10, 15, or 20% (see Chapter 6 for additional detail on this risk metric). Estimates are presented for both 2007 and 2009 within each figure. In order to compare, for example, exposure estimates (based on the 20 percent FEV1 decrement) for the urban core between current conditions and the current standard for 2007, we would compare the light tan column for *urb_122_base_07* with the light tan column for *urb_122_75_07*.

Generally for both the exposure and lung-function risk estimates, we see either a pattern of risk reduction or no change in risk when we look across air quality scenarios (recent conditions – current standard – alternative standard 70 ppb) for a given sub-region (i.e., urban core, outer ring or total combined area). Note however, that in one case (Boston for 2009 for the urban sub-region) we do see a slight risk increase for both exposure and lung function risk (see Figure 9A-3 and 9A-15, respectively). When we compare patterns of risk reduction for the urban core and outer ring (across urban study areas), we generally see larger degrees of risk reduction for the outer rings. This may reflect two factors: (a) design monitors (targeted for ozone reductions under simulated attainment of the current and alternative standard levels) tend to be located in the outer ring and consequently ozone levels near these monitors are likely to experience greater degrees of reduction and (b) there may be a degree of dampening of risk reduction in the urban core reflecting the non-linear nature of ozone formation which can result in increase in ozone on lower ozone days following simulation of both current and alternative standard levels (see section 7.1.1 for additional discussion).

Figure 9A-1 Exposure Risk Estimates – Percent of Person with 1-Hour Exposures at or Above Benchmarks
(Spatially stratified: all study area, urban study area, outer study area) **(Atlanta)**

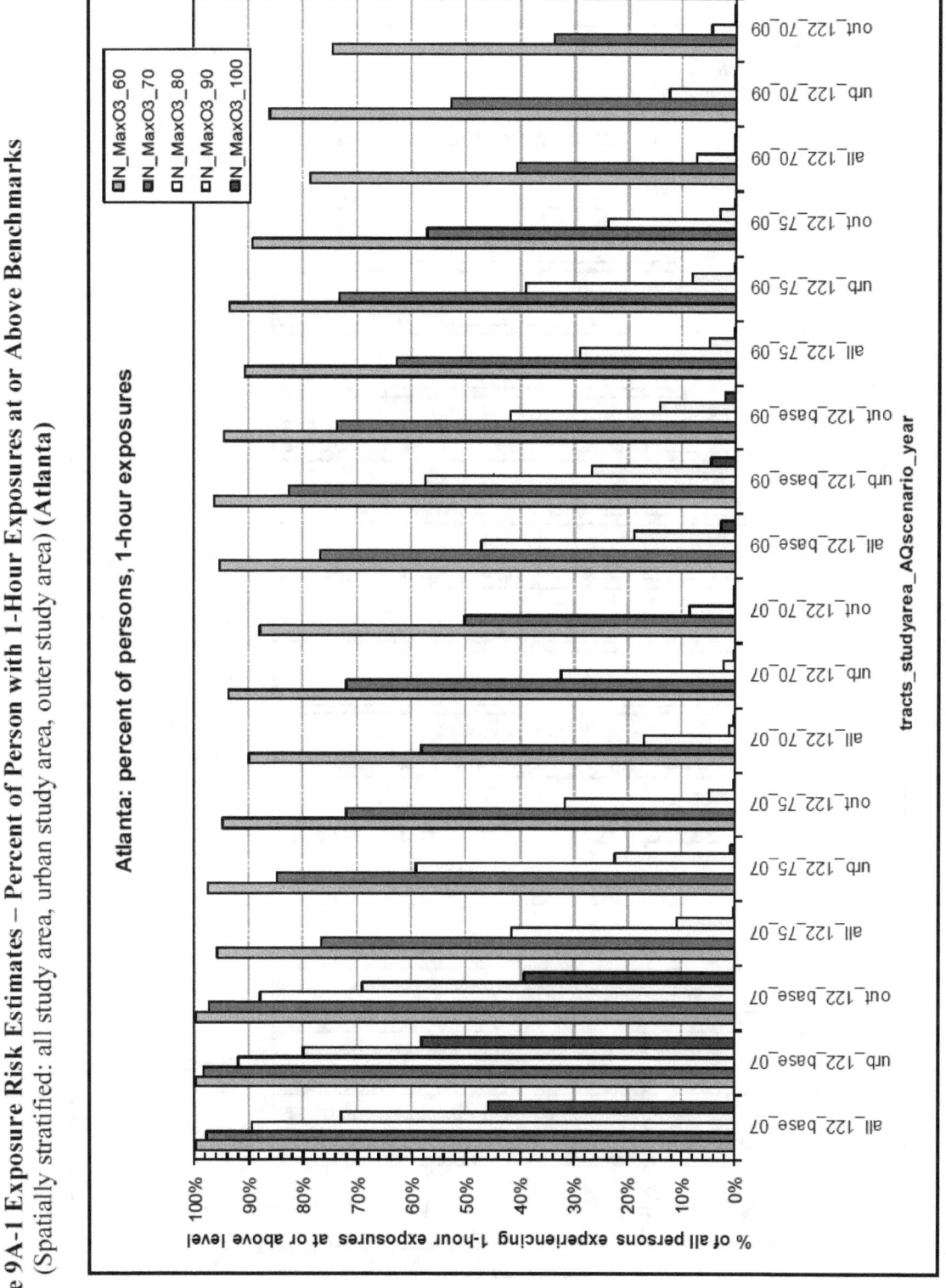

Atlanta: percent of persons, 1-hour exposures

Figure 9A-2 Exposure Risk Estimates – Percent of Person with 1-Hour Exposures at or Above Benchmarks
(Spatially stratified: all study area, urban study area, outer study area) **(Baltimore)**

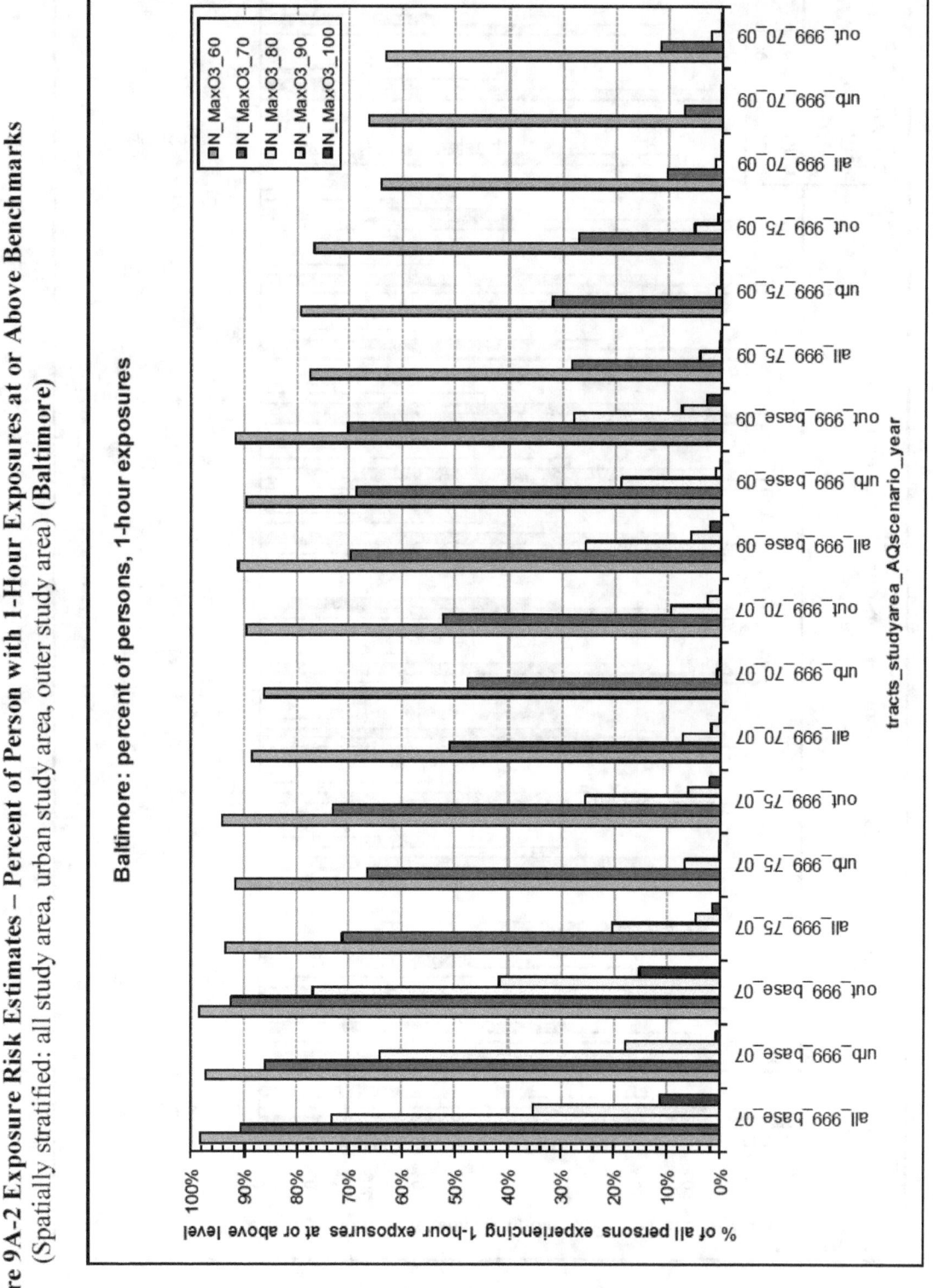

Baltimore: percent of persons, 1-hour exposures

Figure 9A-3 Exposure Risk Estimates – Percent of Person with 1-Hour Exposures at or Above Benchmarks
(Spatially stratified: all study area, urban study area, outer study area) (Boston)

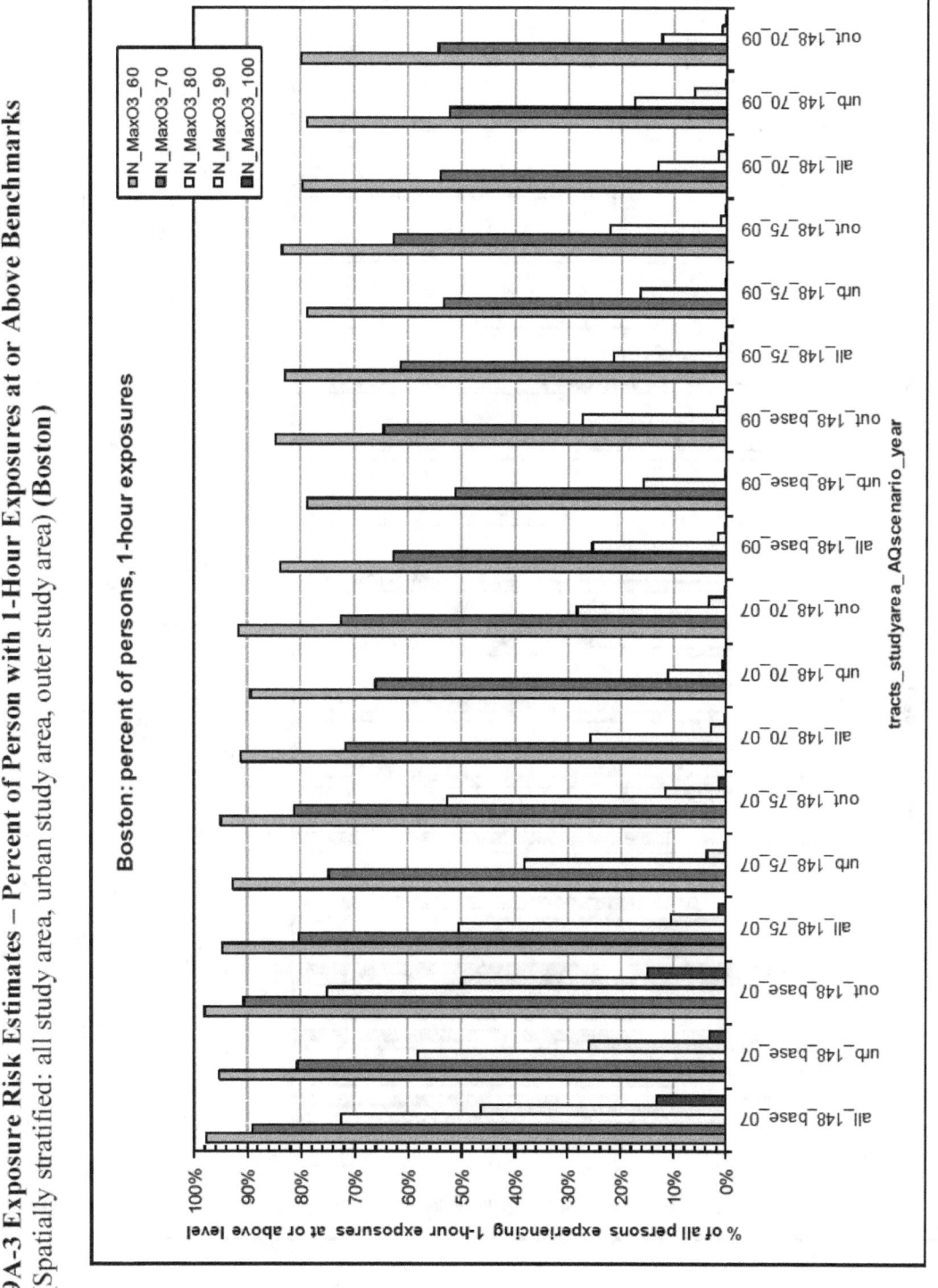

Boston: percent of persons, 1-hour exposures

Figure 9A-4 Exposure Risk Estimates – Percent of Person with 1-Hour Exposures at or Above Benchmarks (Spatially stratified: all study area, urban study area, outer study area) **(Cleveland)**

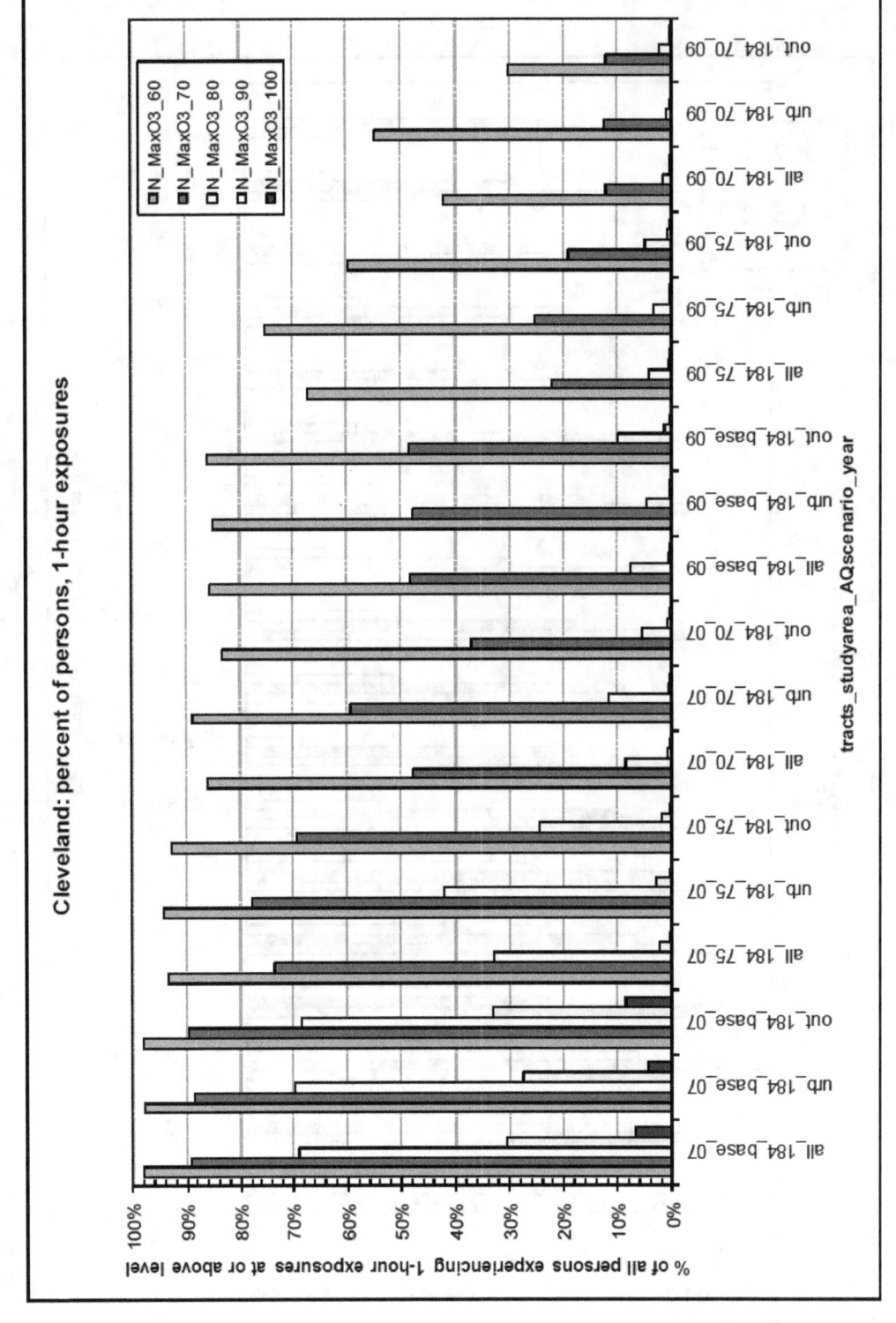

Figure 9A-5 Exposure Risk Estimates – Percent of Person with 1-Hour Exposures at or Above Benchmarks
(Spatially stratified: all study area, urban study area, outer study area) (Denver)

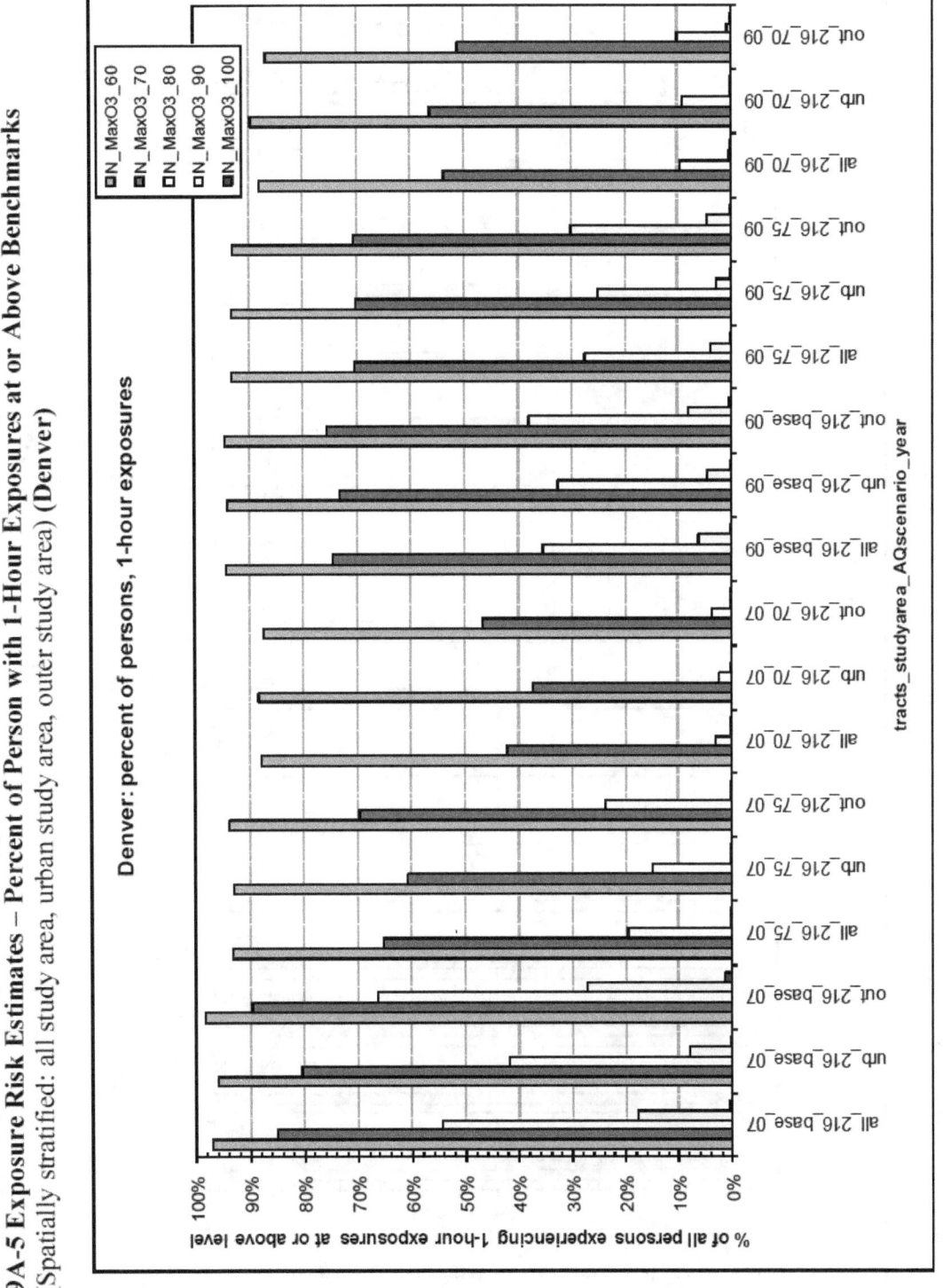

Figure 9A-6 Exposure Risk Estimates – Percent of Person with 1-Hour Exposures at or Above Benchmarks
(Spatially stratified: all study area, urban study area, outer study area) (Detroit)

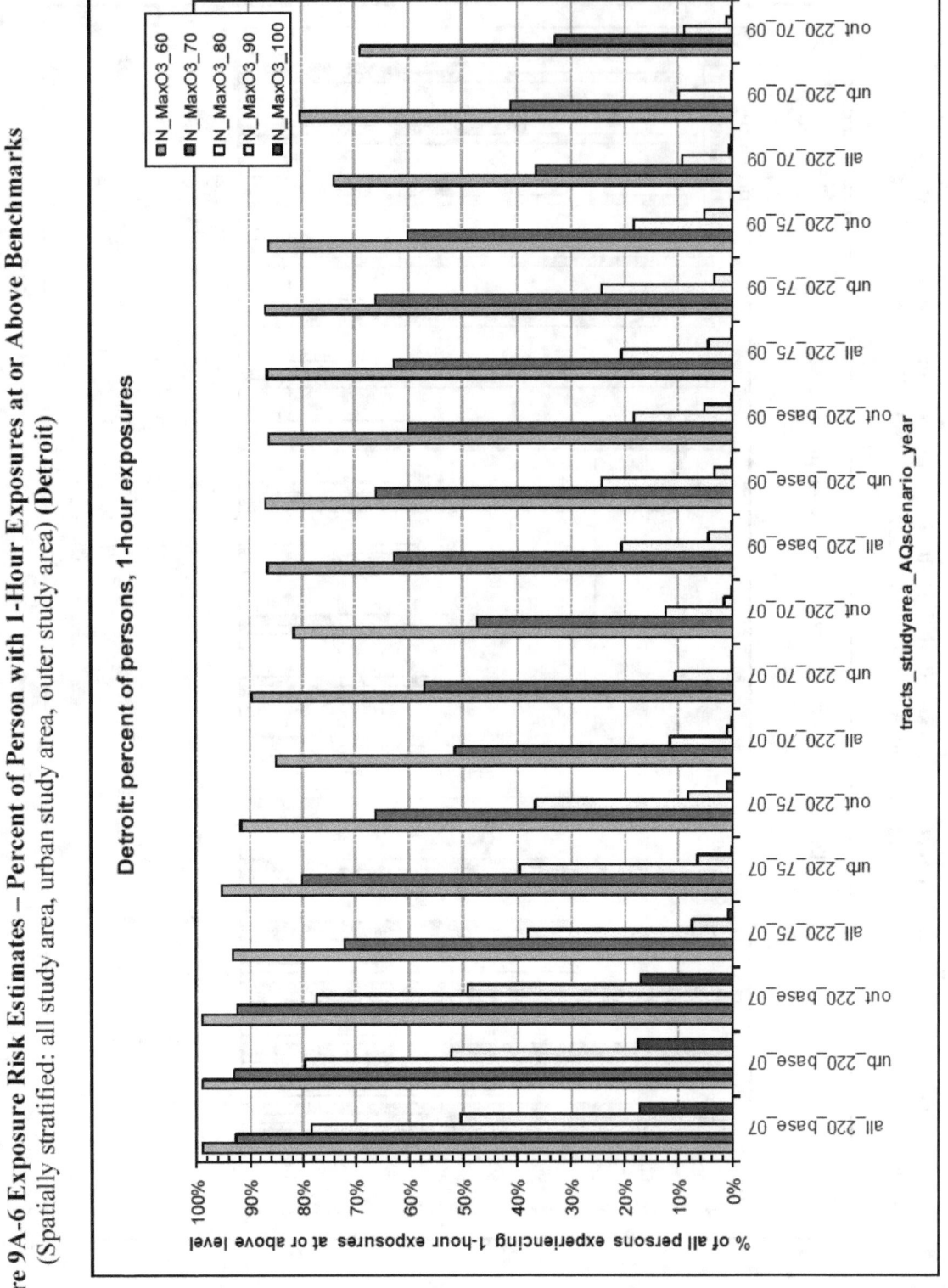

Figure 9A-7 Exposure Risk Estimates – Percent of Person with 1-Hour Exposures at or Above Benchmarks
(Spatially stratified: all study area, urban study area, outer study area) (Houston)

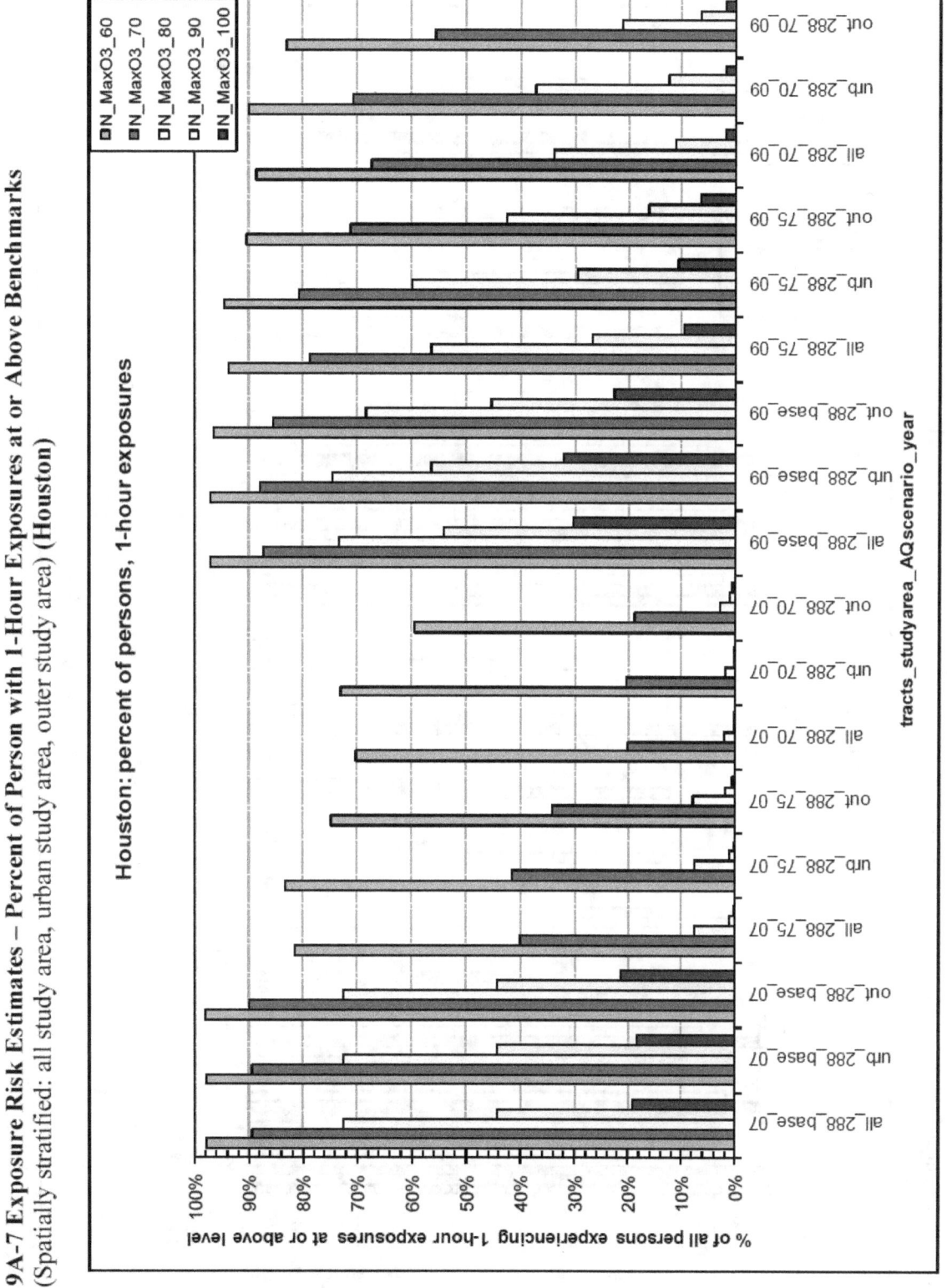

Figure 9A-8 Exposure Risk Estimates – Percent of Person with 1-Hour Exposures at or Above Benchmarks
(Spatially stratified: all study area, urban study area, outer study area) (Los Angeles)

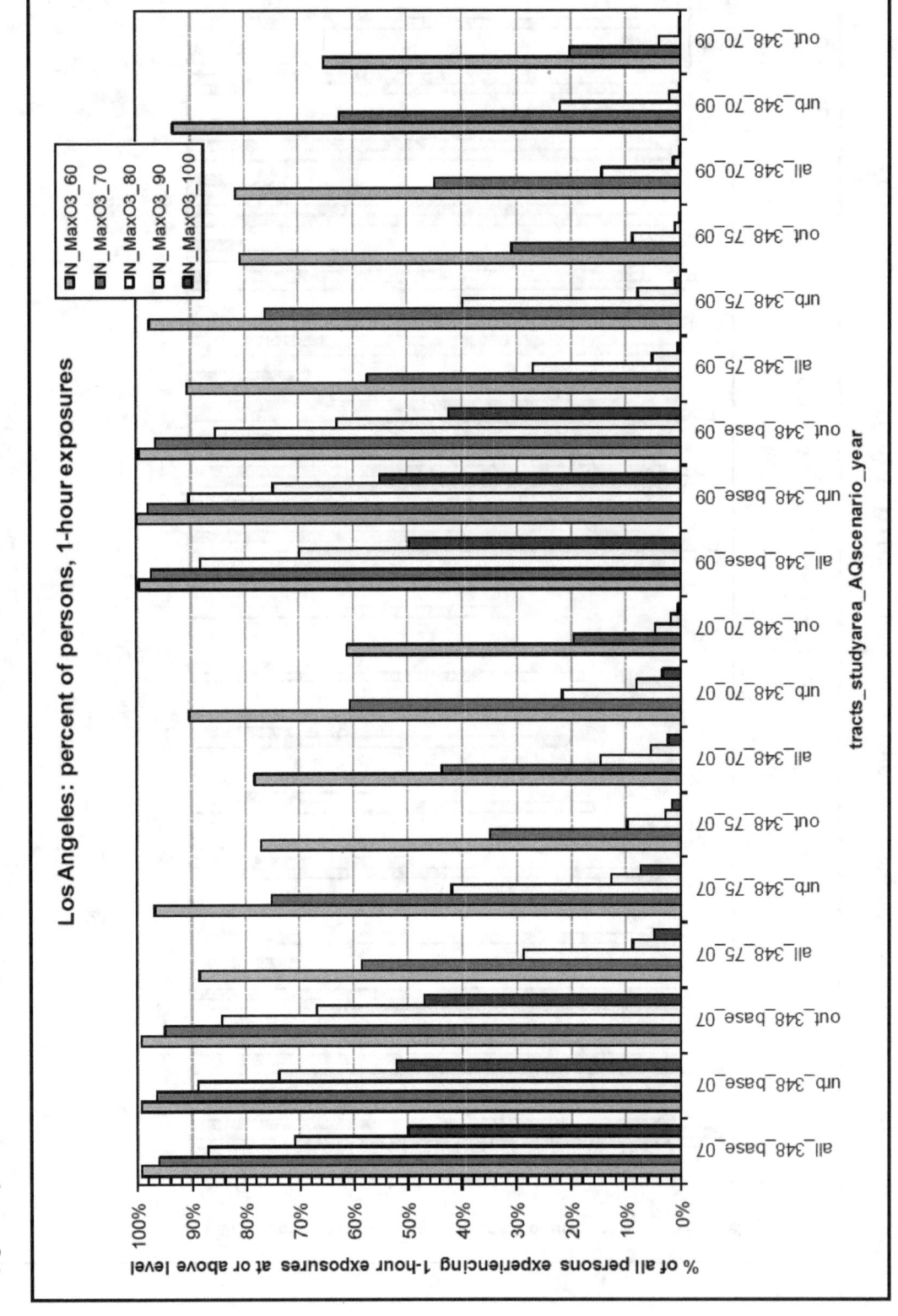

Figure 9A-9 Exposure Risk Estimates – Percent of Person with 1-Hour Exposures at or Above Benchmarks
(Spatially stratified: all study area, urban study area, outer study area) (New York)

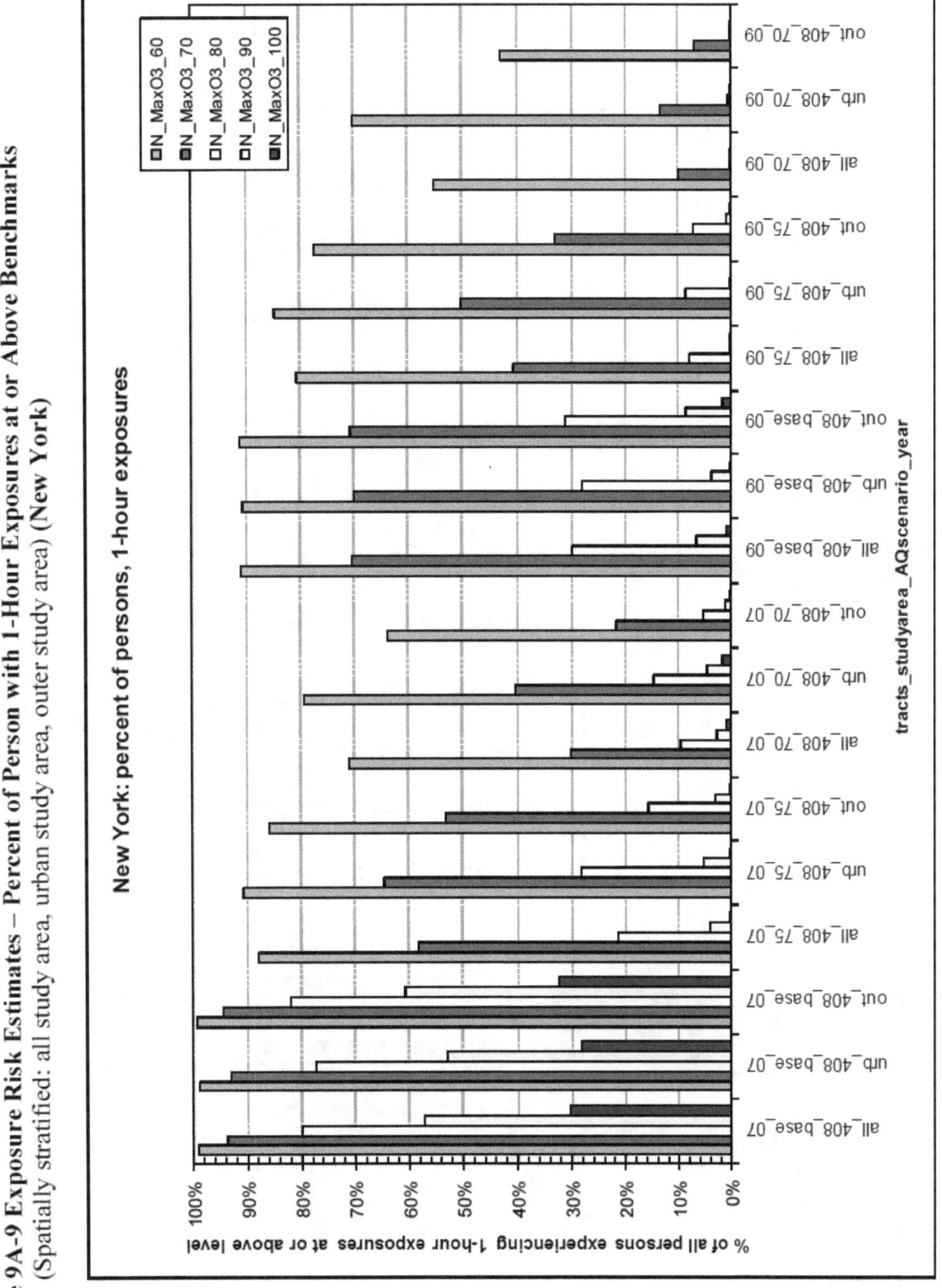

New York: percent of persons, 1-hour exposures

Figure 9A-10 Exposure Risk Estimates – Percent of Person with 1-Hour Exposures at or Above Benchmarks
(Spatially stratified: all study area, urban study area, outer study area) **(Philadelphia)**

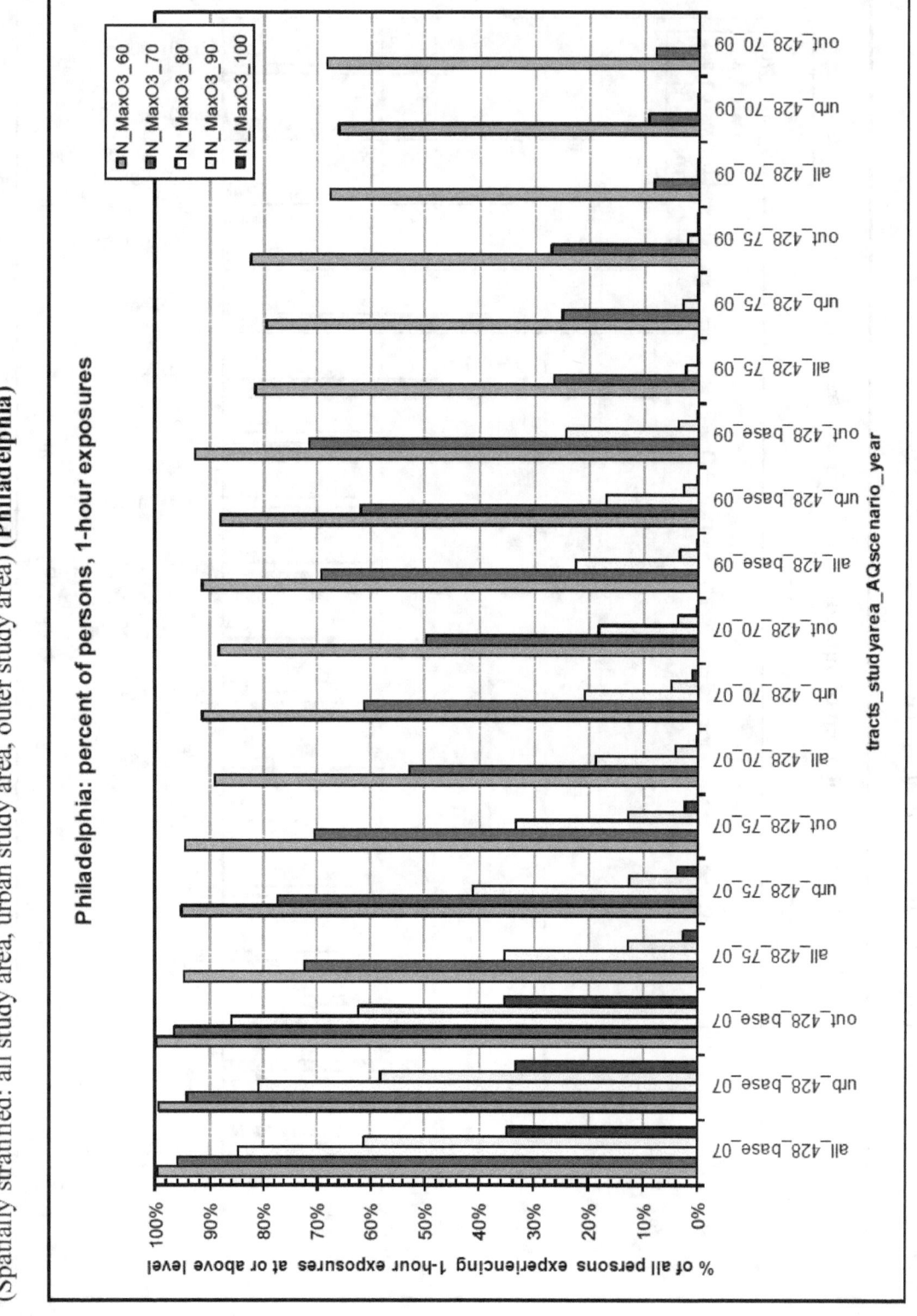

Philadelphia: percent of persons, 1-hour exposures

Figure 9A-11 Exposure Risk Estimates – Percent of Person with 1-Hour Exposures at or Above Benchmarks
(Spatially stratified: all study area, urban study area, outer study area) (Sacramento)

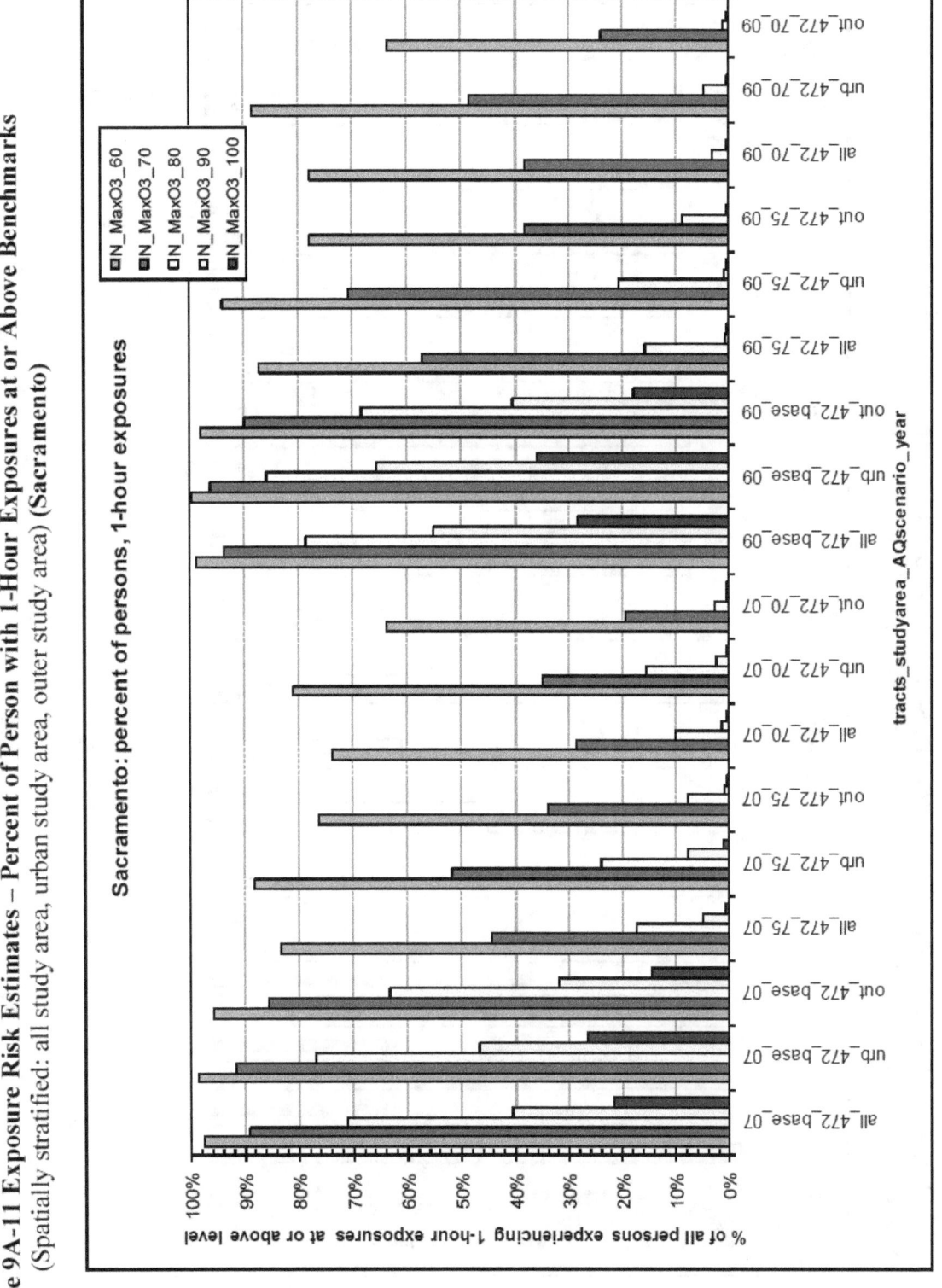

Figure 9A-12 Exposure Risk Estimates – Percent of Person with 1-Hour Exposures at or Above Benchmarks
(Spatially stratified: all study area, urban study area, outer study area) (St. Louis)

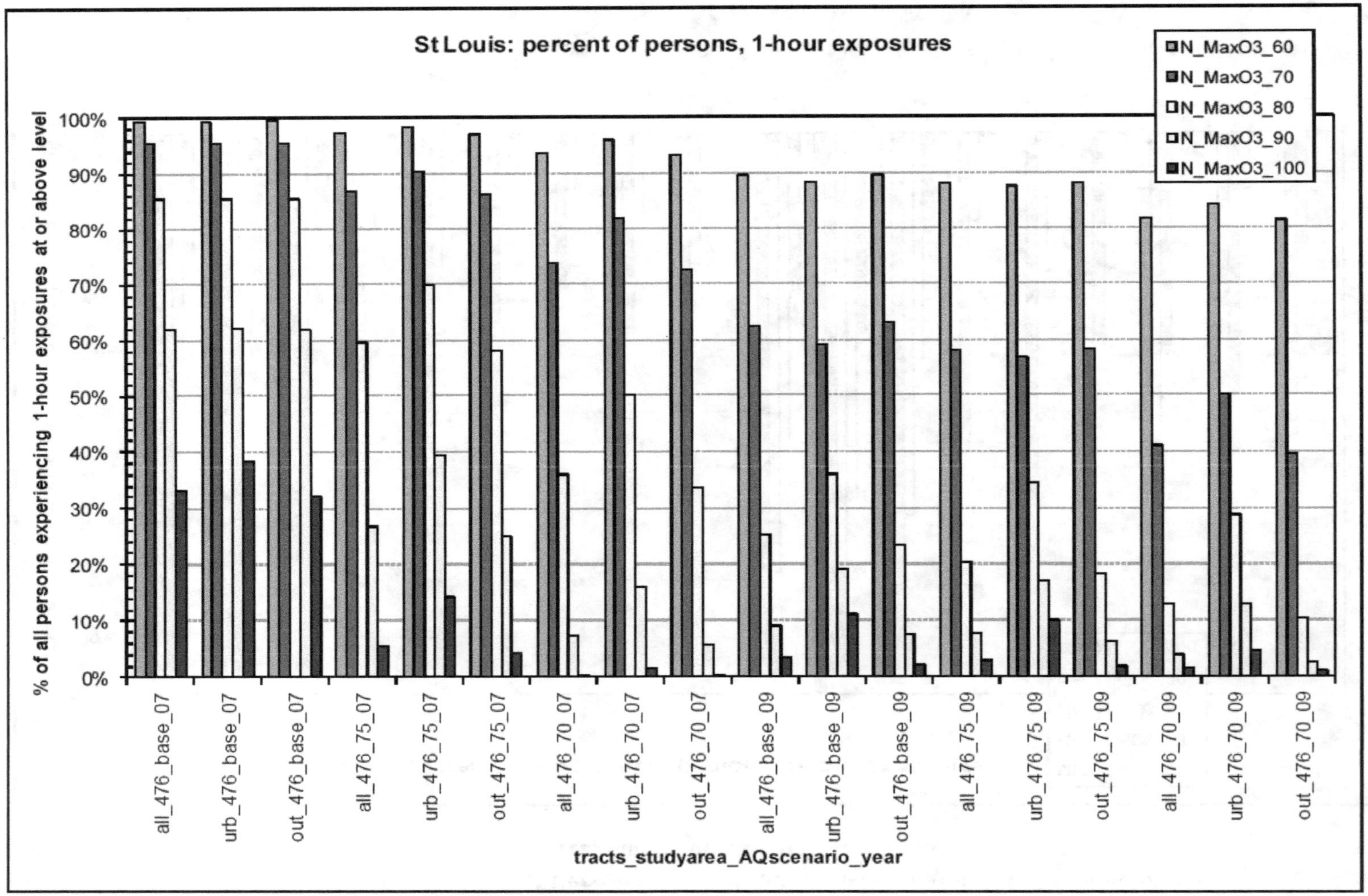

St Louis: percent of persons, 1-hour exposures

Figure 9A-13 Lung-Function Risk Estimates – Percent of Person with Specified FEV1 Decrement
(Spatially stratified: all study area, urban study area, outer study area) (Atlanta)

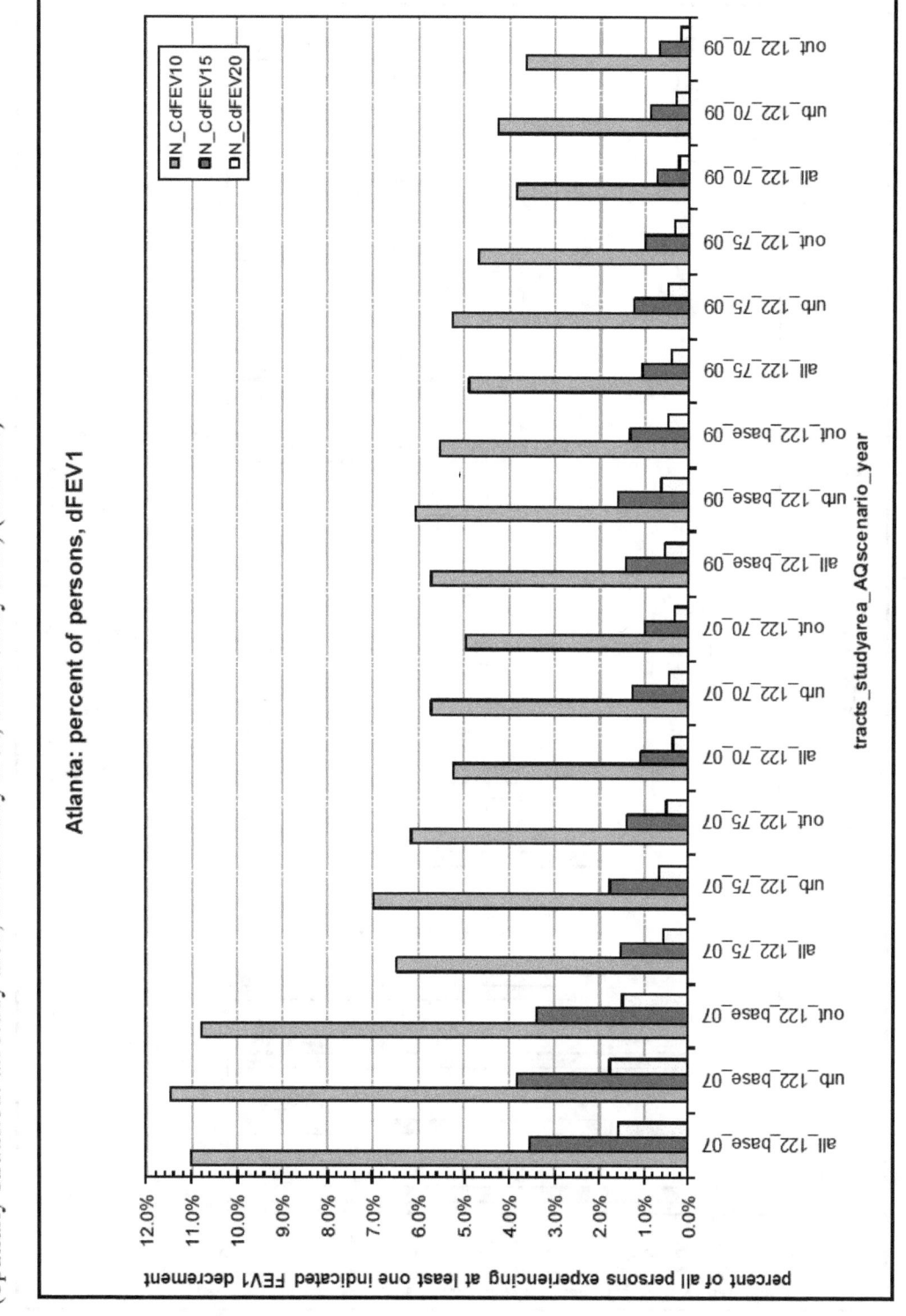

Atlanta: percent of persons, dFEV1

Figure 9A-14 Lung-Function Risk Estimates – Percent of Person with Specified FEV1 Decrement
(Spatially stratified: all study area, urban study area, outer study area) **(Baltimore)**

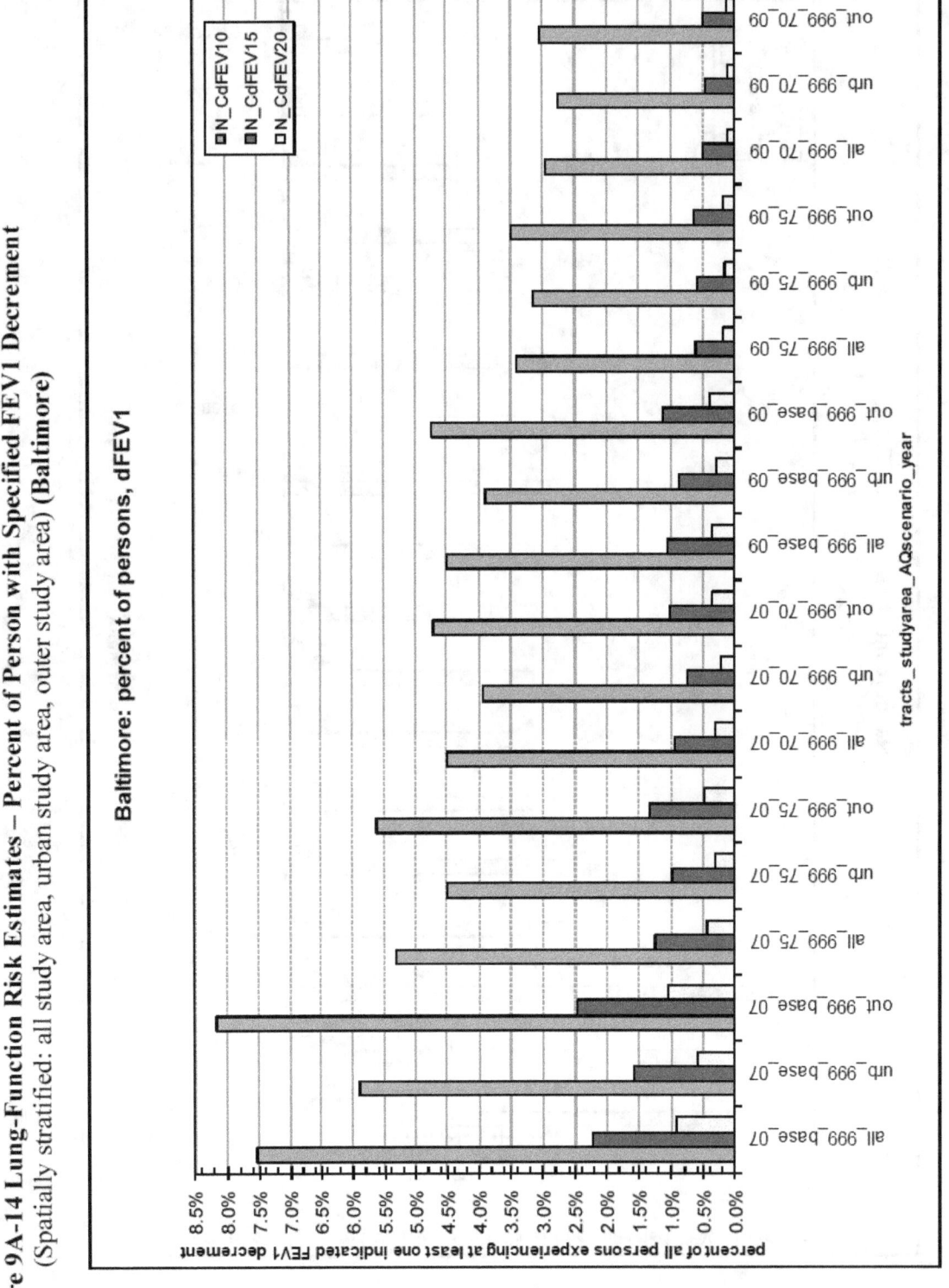

Figure 9A-15 Lung-Function Risk Estimates – Percent of Person with Specified FEV1 Decrement (Spatially stratified: all study area, urban study area, outer study area) (Boston)

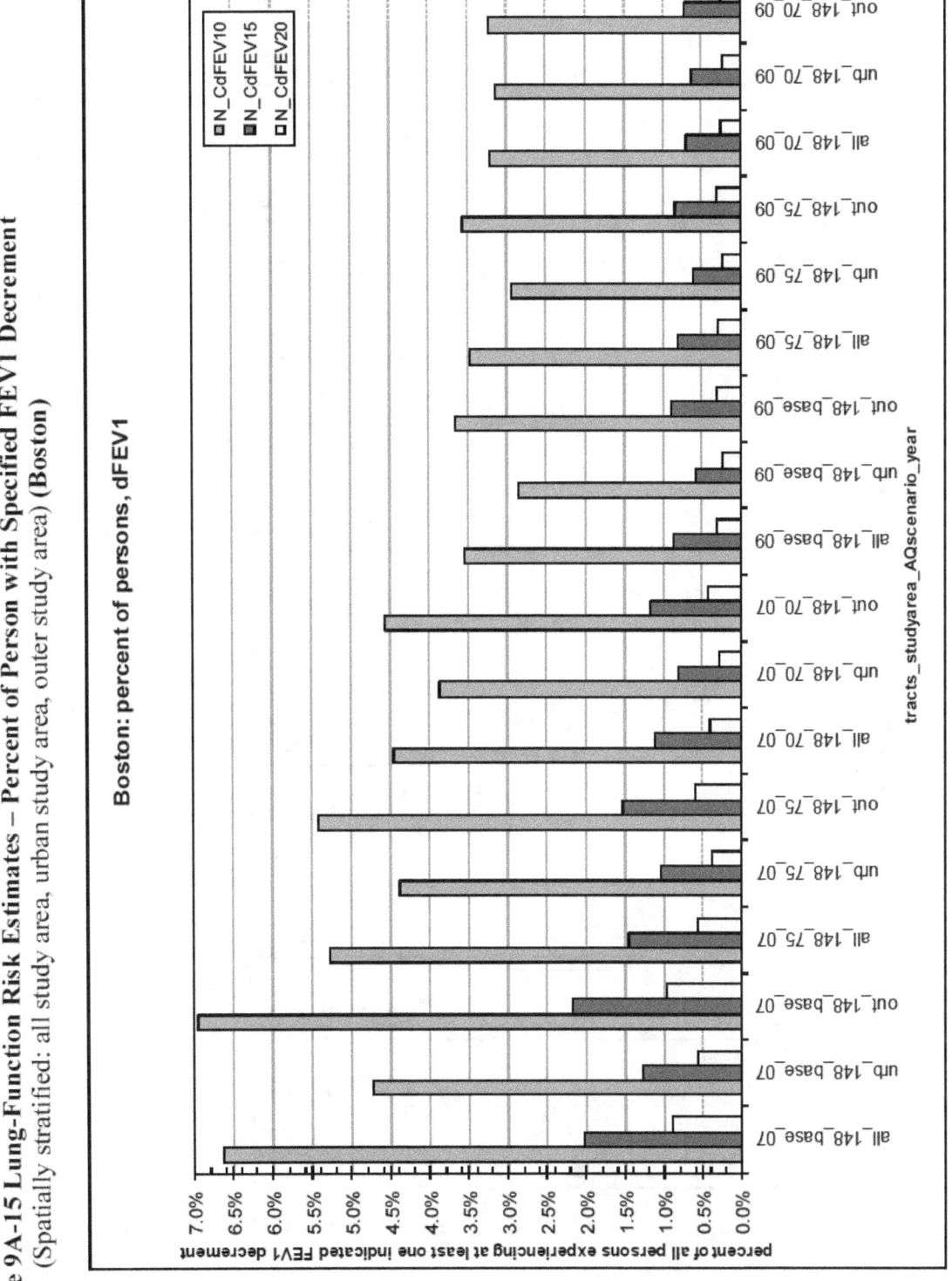

Figure 9A-16 Lung-Function Risk Estimates – Percent of Person with Specified FEV1 Decrement
(Spatially stratified: all study area, urban study area, outer study area) (Cleveland)

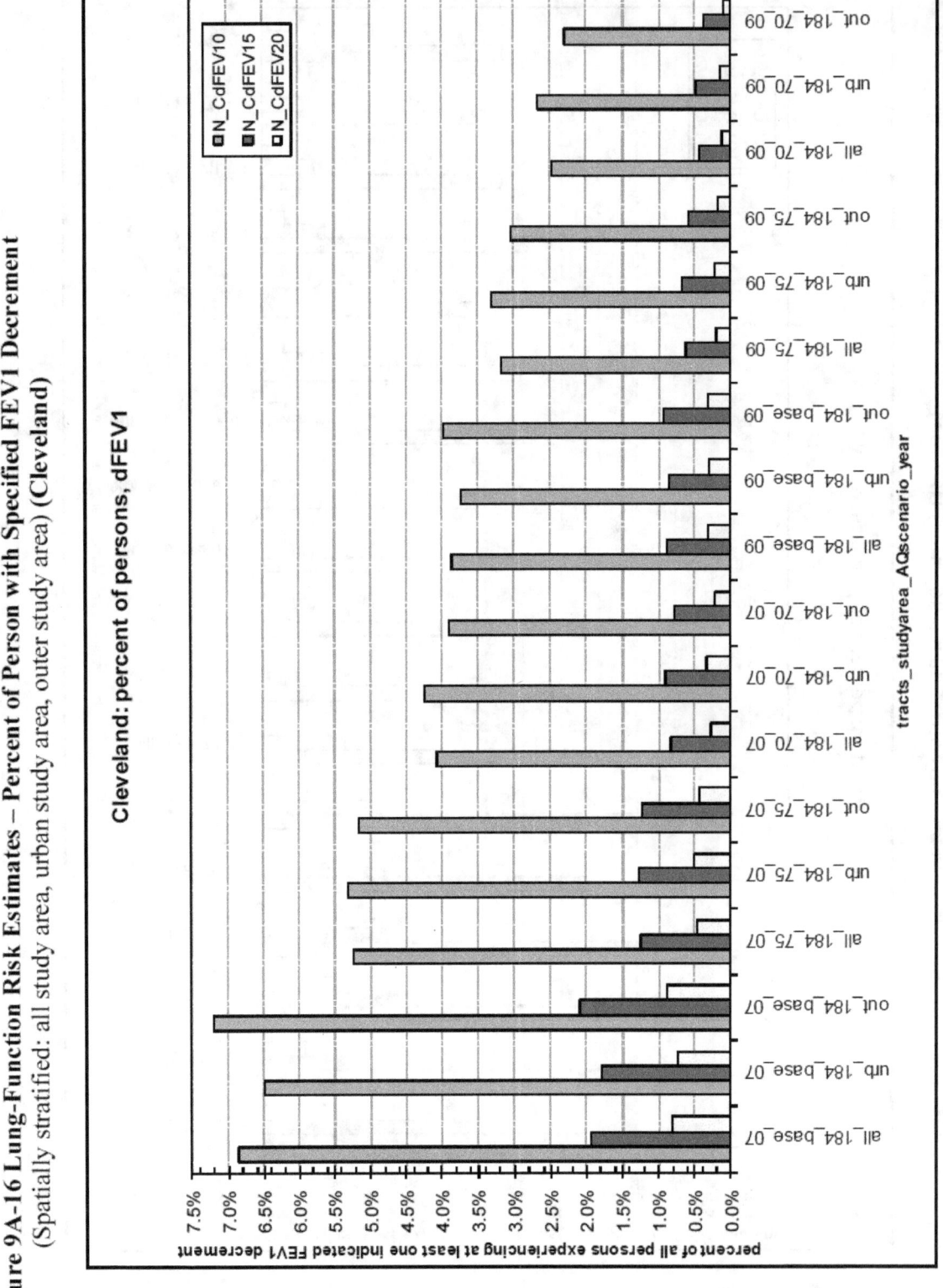

Figure 9A-17 Lung-Function Risk Estimates – Percent of Person with Specified FEV1 Decrement
(Spatially stratified: all study area, urban study area, outer study area) (Denver)

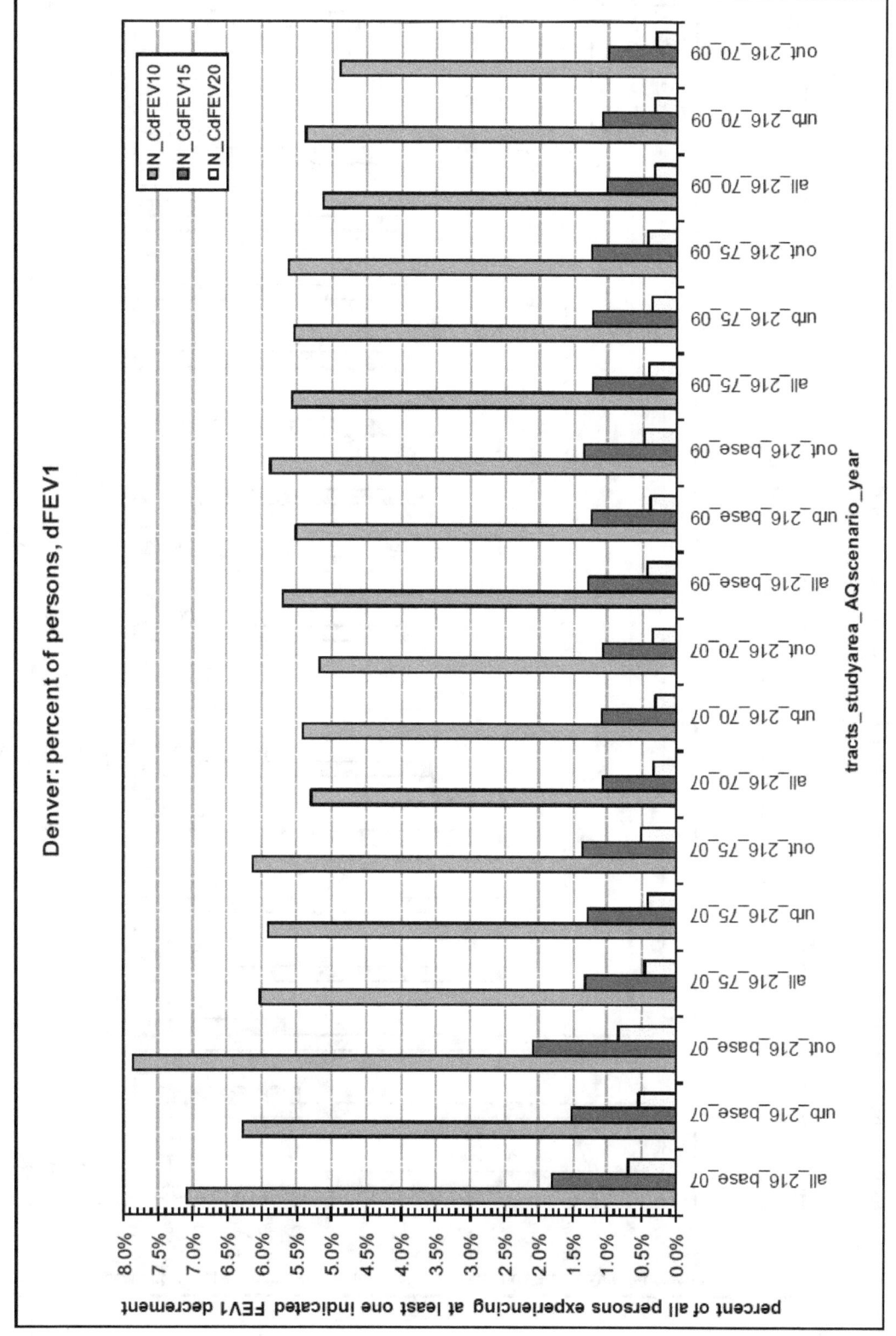

Figure 9A-18 Lung-Function Risk Estimates – Percent of Person with Specified FEV1 Decrement (Spatially stratified: all study area, urban study area, outer study area) (Detroit)

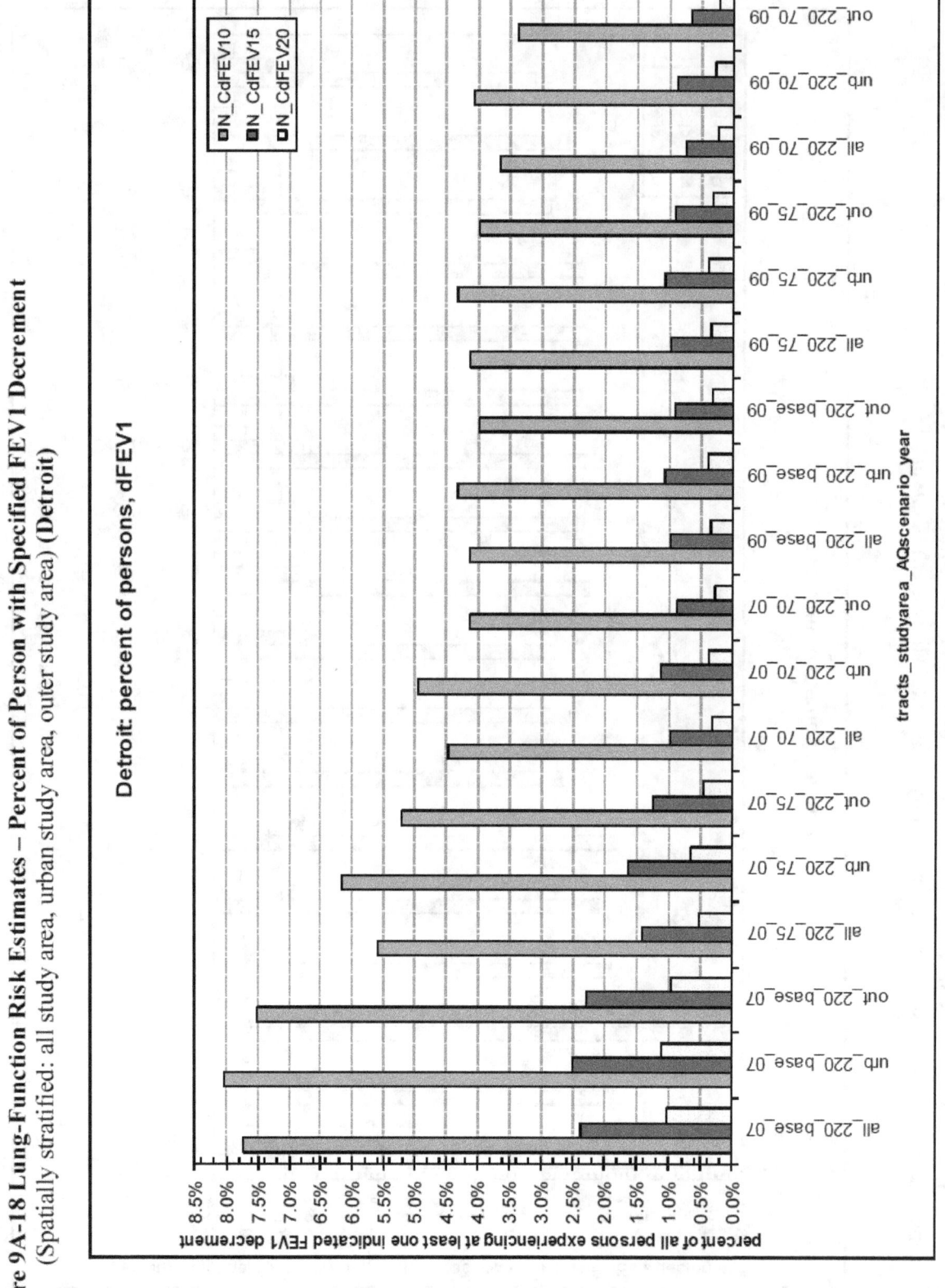

Figure 9A-19 Lung-Function Risk Estimates – Percent of Person with Specified FEV1 Decrement
(Spatially stratified: all study area, urban study area, outer study area) **(Houston)**

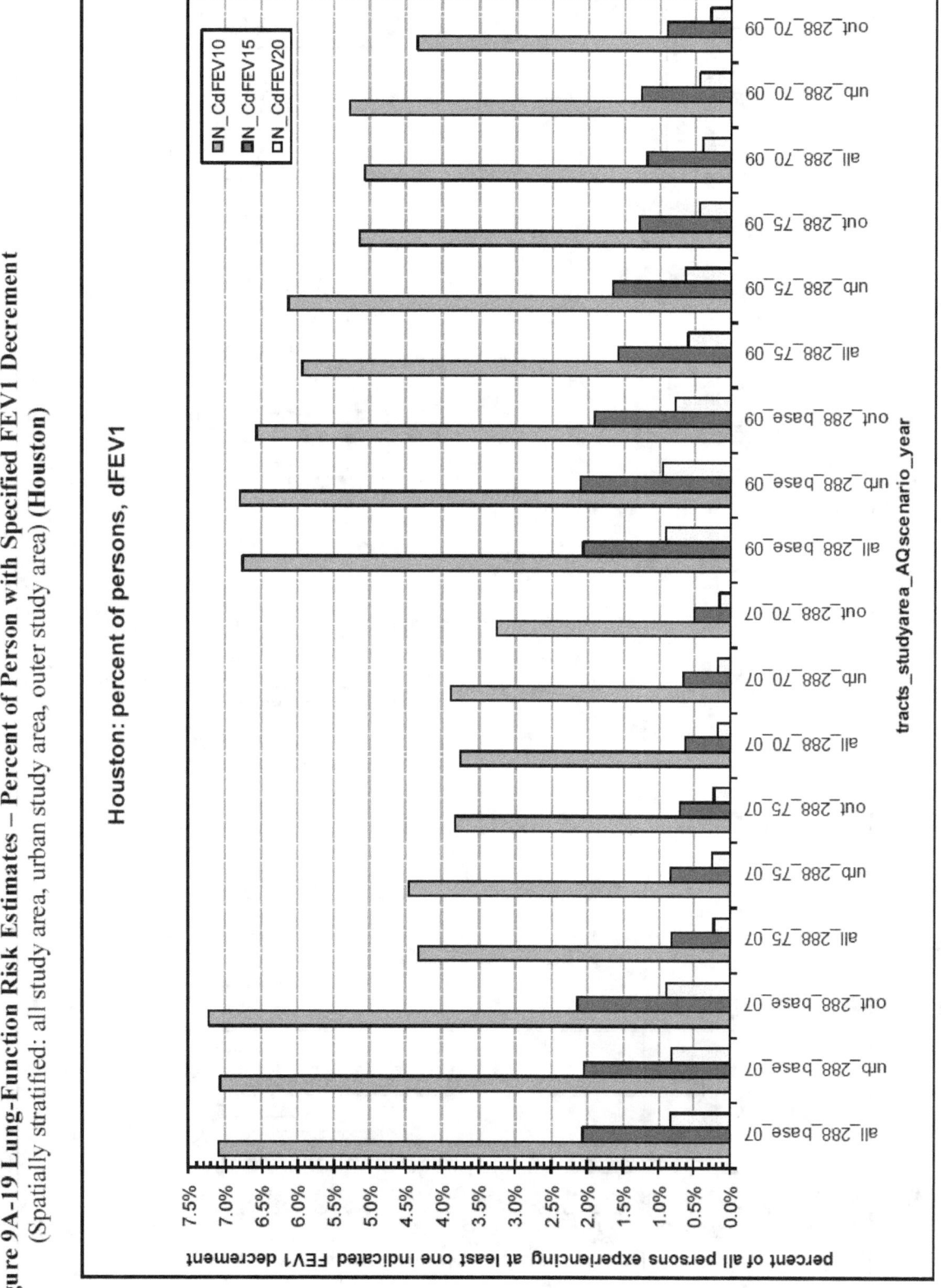

Houston: percent of persons, dFEV1

Figure 9A-20 Lung-Function Risk Estimates – Percent of Person with Specified FEV1 Decrement
(Spatially stratified: all study area, urban study area, outer study area) (Los Angeles)

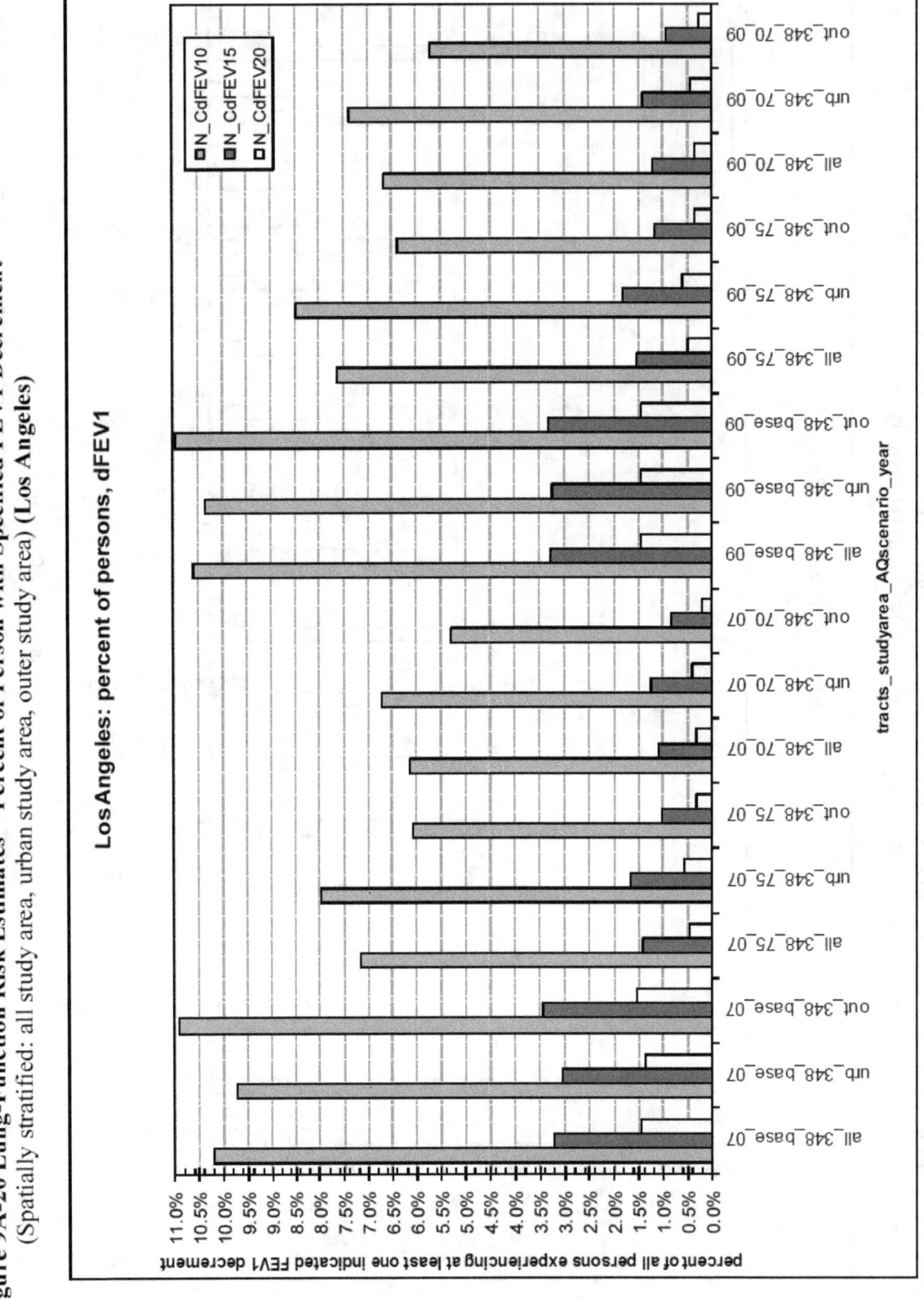

Los Angeles: percent of persons, dFEV1

Figure 9A-21 Lung-Function Risk Estimates – Percent of Person with Specified FEV1 Decrement
(Spatially stratified: all study area, urban study area, outer study area) (New York)

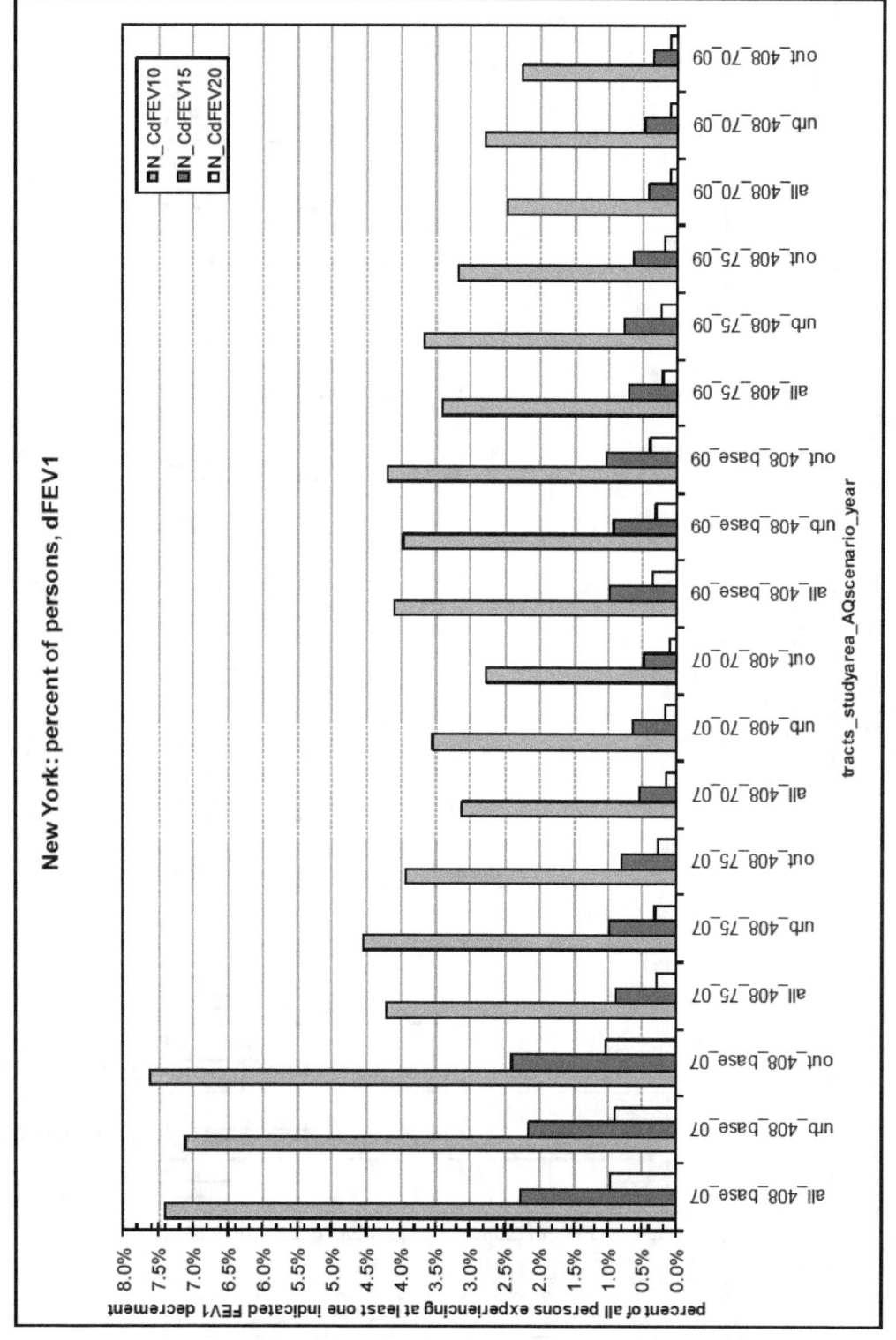

Figure 9A-22 Lung-Function Risk Estimates – Percent of Person with Specified FEV1 Decrement (Spatially stratified: all study area, urban study area, outer study area) (Philadelphia)

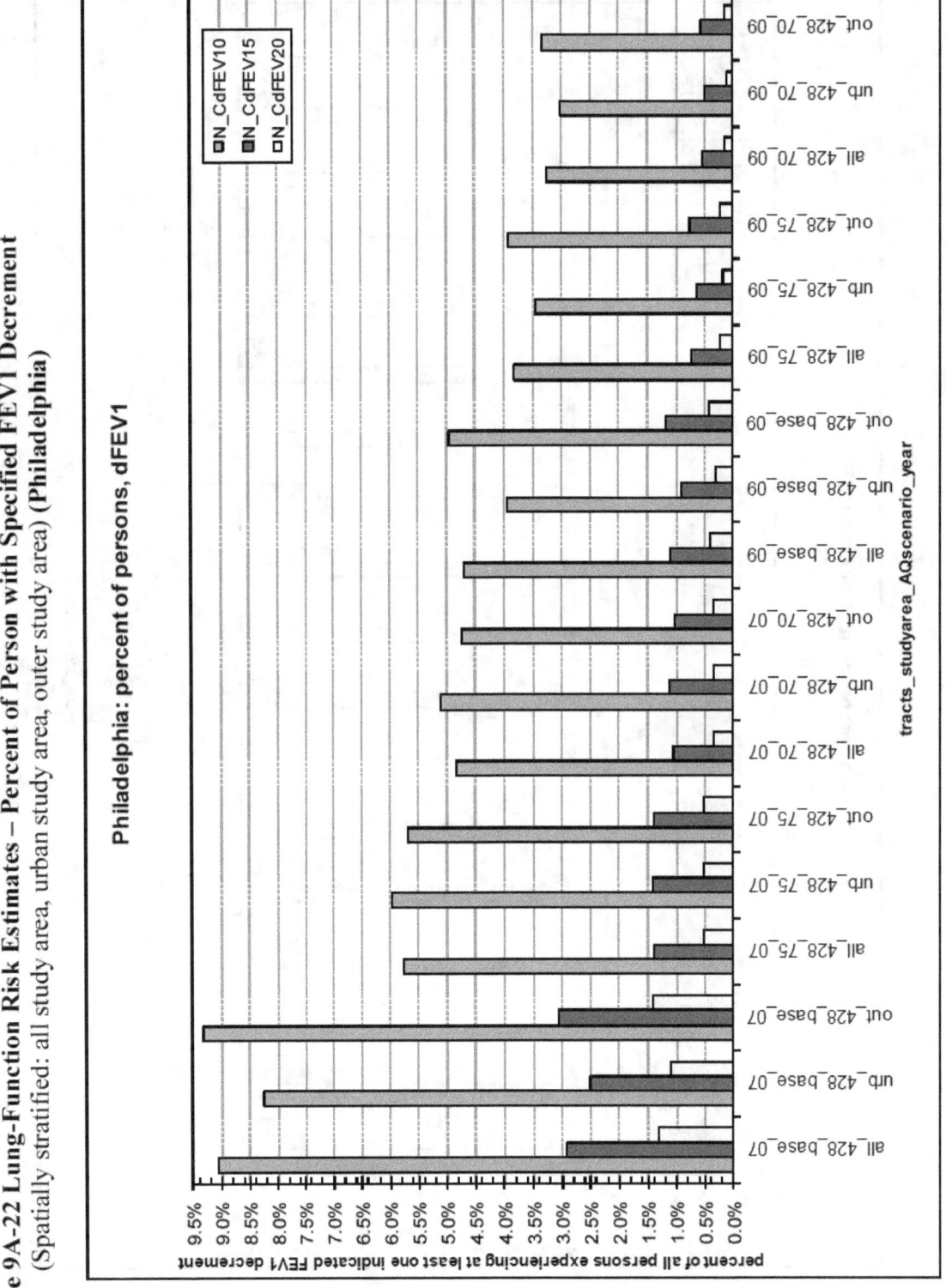

Figure 9A-23 Lung-Function Risk Estimates – Percent of Person with Specified FEV1 Decrement
(Spatially stratified: all study area, urban study area, outer study area) (Sacramento)

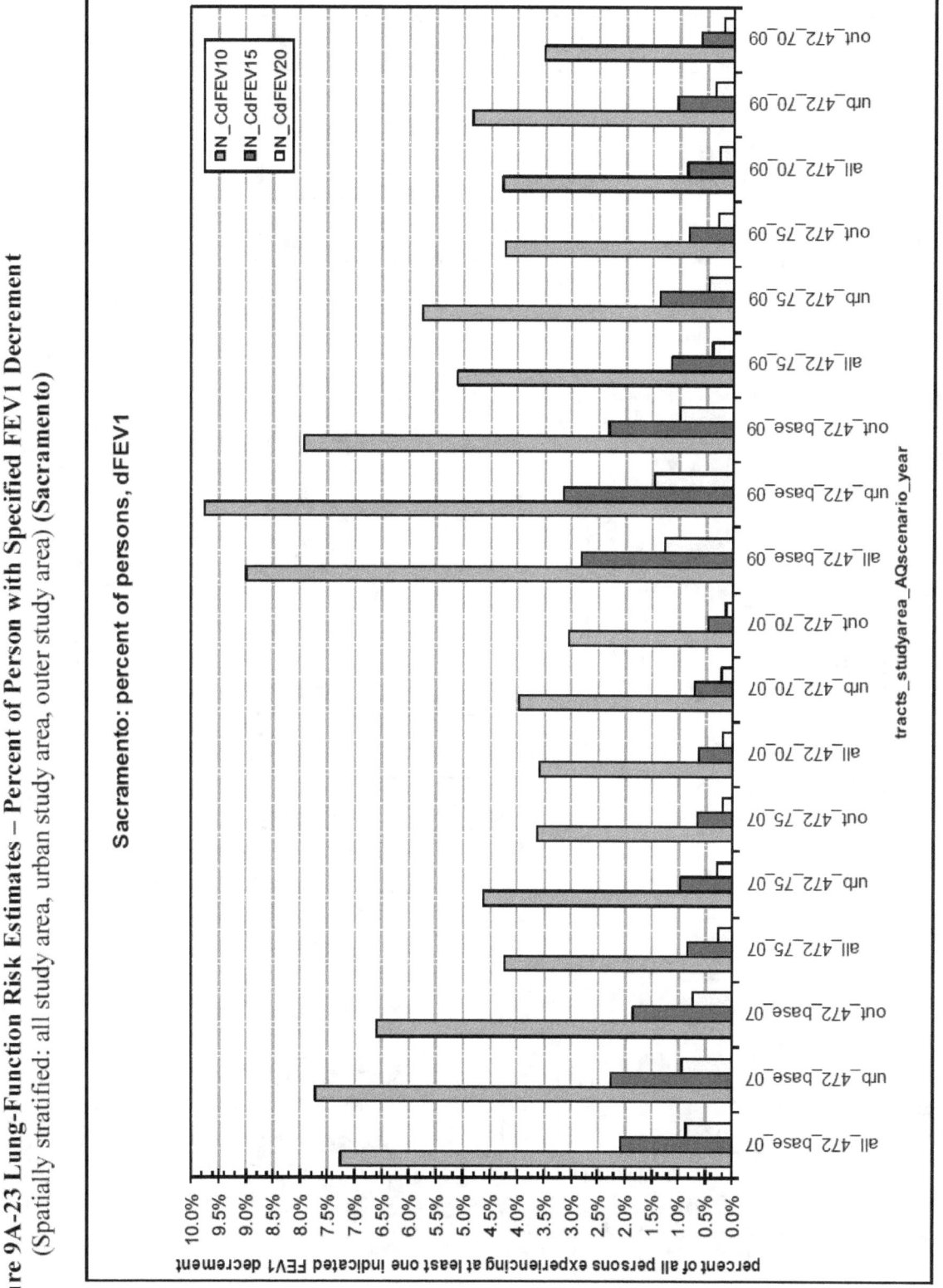

Figure 9A-24 Lung-Function Risk Estimates – Percent of Person with Specified FEV1 Decrement
(Spatially stratified: all study area, urban study area, outer study area) (St Louis)

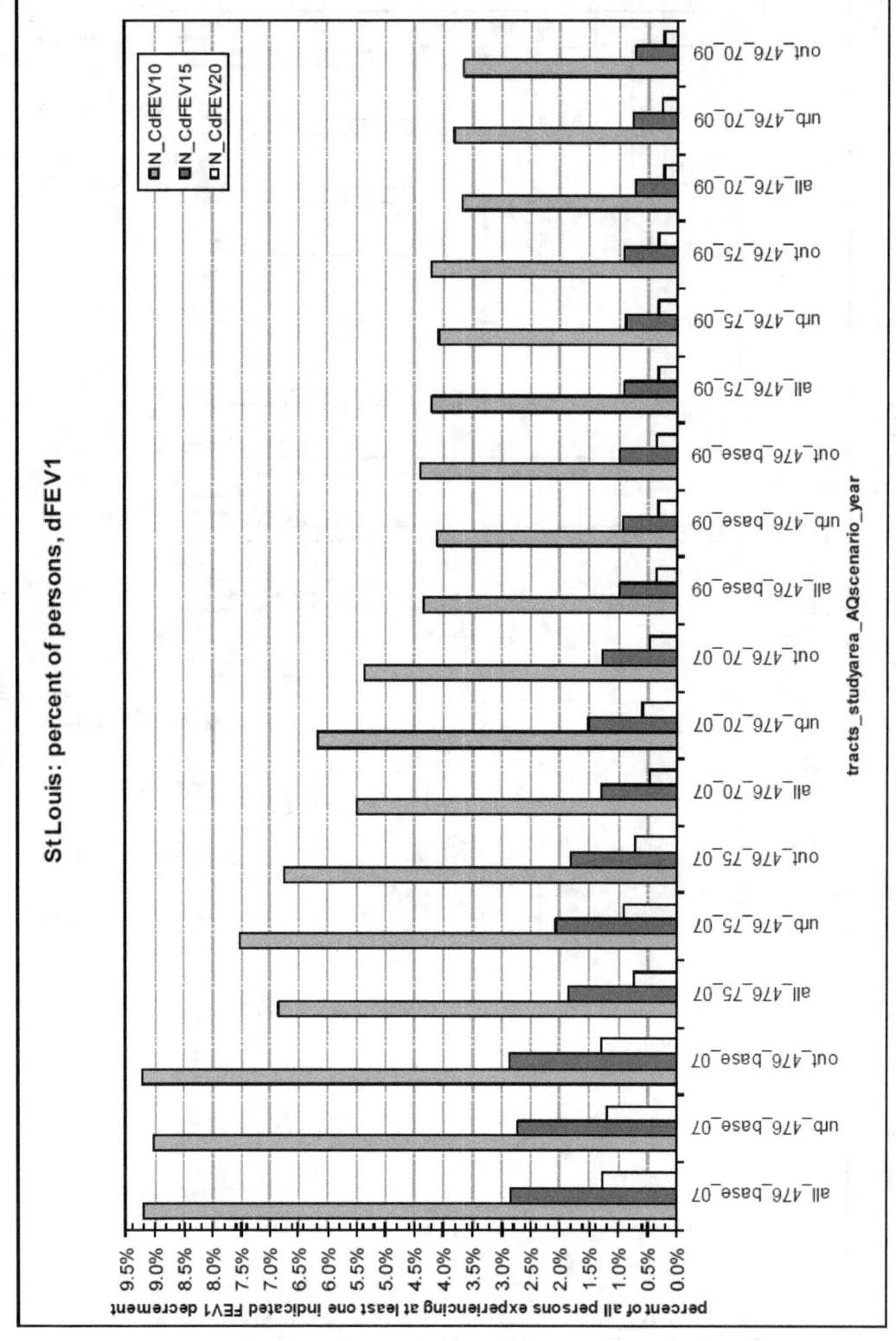

United States
Environmental Protection
Agency

Office of Air Quality Planning and Standards
Air Quality Strategies and Standards Division
Research Triangle Park, NC

Publication No. EPA-452/P-14-004e
February 2014

www.ingramcontent.com/pod-product-compliance
Lightning Source LLC
Chambersburg PA
CBHW081451170526
45166CB00008B/2386